1 MONTH OF
FREE
READING

at

www.ForgottenBooks.com

By purchasing this book you are eligible for one month membership to ForgottenBooks.com, giving you unlimited access to our entire collection of over 1,000,000 titles via our web site and mobile apps.

To claim your free month visit:
www.forgottenbooks.com/free909072

ISBN 978-0-265-91482-3
PIBN 10909072

Forgotten Books is a registered trademark of FB &c Ltd.
Copyright © 2018 FB &c Ltd.
FB &c Ltd, Dalton House, 60 Windsor Avenue, London, SW19 2RR.
Company number 08720141. Registered in England and Wales.

For support please visit www.forgottenbooks.com

ARITHMETIC

G. M. Lander

UPON THE

INDUCTIVE METHOD OF INSTRUCTION:

BEING

A SEQUEL

TO

INTELLECTUAL ARITHMETIC.

—◆—

BY WARREN COLBURN, A. M.

—◆—

STEREOTYPED AT THE BOSTON TYPE AND STEREOTYPE FOUNDRY.

BOSTON:

HILLIARD, GRAY, LITTLE, AND WILKINS.

1828.

DISTRICT OF MASSACHUSETTS, to wit:

District Clerk's Office.

BE IT REMEMBERED, that on the twenty-fifth day of May, A. D. 1826, and in the fiftieth year of the Independence of the United States of America, Warren Colburn, of the said district, has deposited in this office the title of a book, the right whereof he claims as author, in the words following, to wit:

"Arithmetic upon the Inductive Method of Instruction: being a Sequel to Intellectual Arithmetic. By Warren Colburn, A. M."

In conformity to the Act of the Congress of the United States, entitled, "An act for the encouragement of learning, by securing the copies of maps, charts, and books, to the authors and proprietors of such copies, during the times therein mentioned;" and also an act, entitled, "An act supplementary to an act, entitled, An act for the encouragement of learning, by securing the copies of maps, charts, and books, to the authors and proprietors of such copies, during the times therein mentioned; and extending the benefits thereof to the arts of designing, engraving, and etching, historical and other prints"

JNO. W. DAVIS,
Clerk of the District of Massachusetts.

RECOMMENDATIONS.

From B. A. GOULD, Principal of the Public Latin School, Boston.

Boston, 22d Oct., 1822.

DEAR SIR,

I have been highly gratified by the examination of the second part of your Arithmetic. The principles of the science are unfolded, and its practical uses explained with great perspicuity and simplicity. I think your reasonings and illustrations are peculiarly happy and original. This, together with your "First Lessons," forms the most lucid and intelligible, as well as the most scientific system of Arithmetic I have ever seen.—Its own merits place it beyond the need of commendation.

With much esteem,

Sir, your obedient servant,

B. A. GOULD.

Mr. WARREN COLBURN.

From G. B. EMERSON, Principal of the English Classical School, Boston.

Boston, 22d Oct., 1822.

DEAR SIR,

I have carefully examined a large portion of your manuscript, and do not hesitate to recommend it very highly to every person who wishes to teach arithmetic intelligibly. The arrangement is very much better, the explanations more convincing, and the rules, from the mode in which they are introduced, are clearer and simpler, than can be found in any book on the subject with which I am acquainted.

I am, with great respect,

Yours, &c.

G. B. EMERSON.

Mr. WARREN COLBURN.

PREFACE.

It will be extremely useful, though not absolutely necessary, for pupils of every age to study the "First Lessons," previous to commencing this treatise. There is an intimate connexion between the two, though this is not dependent on the other. It is hoped that this will be found less difficult than other treatises on the subject, for those who have not studied the "First Lessons."

Pupils may commence the "First Lessons" to advantage, as soon as they can read the examples; and even before they can read, it will be found very useful to ask them questions from it. This may be done by other pupils who have already studied it. Those who commence early, may generally obtain sufficient knowledge of it by the time they are eight or nine years old. They may then commence this.

. This Sequel consists of two parts. The first contains a course of examples for the illustration and application of the principles. The second part contains a developement of the principles. The articles are numbered in the two, so as to correspond with each other. The two parts are to be studied together, when the pupil is old enough to comprehend the second part by reading it himself. When he has performed all the examples in an article in the first part, he should be required to recite the corresponding article in the second part, not verbatim, but to give a good account of the reasoning. When the principle is well understood, the rules which are printed in Italics should be committed to memory. At each recitation, the first thing should be to require the pupil to give a practical example, involving the principle to be explained, and then an explanation of the principle itself.

When the pupil is to learn the use of figures for the first time, it is best to explain to him the nature of them as in Art. I., to about three or four places; and then require him to write some numbers. Then give him some of the first examples in Art. II., without telling him what to do. He will discover what is to be done, and invent a way to do it. Let him perform several in his own way, and then suggest some method a little different from his, and nearer the common method. If he readily comprehends it, he will be pleased with it, and adopt it. If he does not, his mind is not yet prepared for it, and should be allowed to continue his own way longer, and then it should be suggested again. After he is familiar with that, suggest another method, somewhat nearer the common method, and so on, until he learns the best method. Never urge him to adopt any method until he understands it, and is pleased with it. In some of the articles, it may perhaps be necessary for young pupils to perform more examples than are given in the book.

When the pupil is to commence multiplication, give him one of the first examples in Art. III., as if it were an example in Addition. He will write it down as such. But if he is familiar with the " First Lessons," he will probably perform it as multiplication without knowing it. When he does this, suggest to him, that he need not write the number but once. Afterwards recommend to him to write a number to show how many times he repeated it, lest he should forget it Then tell him that it is Multiplication. Proceed in a similar manner with the other rules.

One general maxim to be observed with pupils of every age, is never to tell them directly how to perform any example. If a pupil is unable to perform an example, it is generally because he does not fully comprehend the object of it. The object should be explained, and some questions asked, which will have a tendency to recal the principles necessary. If this does not succeed, his mind is not prepared for it, and he must be required to examine it more by himself, and to review some of the principles which it involves. It is useless for him to perform it before his mind is prepared for it. After he has been told, he is satisfied, and will not be willing to examine the principle, and he will be no better prepared for another case of the same kind, than he was before. When the pupil knows that he is not to be told, he learns to depend on himself; and when he once contracts the habit of understanding what he does, he will not easily be prevailed on to do any thing which he does not understand.

Several considerations induce the author to think, that when a principle is to be taught, practical questions should first be proposed, care being taken to select such as will show the combination in the simplest manner, and that the numbers be so small that the operation shall not be difficult. When a proper idea is formed of the nature and use of the combination, the method of solving these questions with large numbers should be attended to. This method, on trial has succeeded beyond his expectations. Practical examples not only show at once the object to be accomplished, but they greatly assist

1 *

the imagination in unfolding the principle and discovering the opera
tions requisite for the solution.

This principle is made the basis of this treatise ; viz. whenever a
new combination is introduced, it is done with practical examples,
proposed in such a manner as to show what it is, and as much as
possible, how it is to be performed. The examples are so small that
the pupil may easily reason upon them, and that there will be no
difficulty in the operation itself, until the combination is well understood. In this way it is believed that the leading idea which the
pupil will obtain of each combination, will be the effect which will be
produced by it, rather than how to perform it, though the latter will
be sufficiently well understood.

The second part contains an analytical developement of the principles. Almost all the examples used for this purpose are practical.
Care has been taken to make every principle depend as little as
possible upon others. Young persons cannot well follow a course of
reasoning where one principle is built upon another. Besides, a principle is always less understood by every one, in proportion as it is
made to depend on others.

In tracing the principles, several distinctions have been made which
have not generally been made. They are principally in division of
whole numbers, and in division of whole numbers by fractions, and
fractions by fractions. There are some instances also of combinations
being classed together, which others have kept separate.

As the purpose is to give the learner a knowledge of the principles,
it is necessary to have the variety of examples under each principle
as great as possible. The usual method of arrangement, according to
subjects, has been on this account entirely rejected, and the arrangement has been made according to principles. Many different subjects
come under the same principle ; and different parts of the same subject frequently come under different principles. When the principles
are well understood, very few subjects will require a particular rule,
and if the pupil is properly introduced to them, he will understand
them better without a rule than with one. Besides, he will be better
prepared for the cases which occur in business, as he will be obliged
to meet them there without a name. The different subjects, as they
are generally arranged, often embarrass the learner. When he meets
with a name with which he is not acquainted, and a rule attached to
it, he is frequently at a loss, when if he saw the example without the
name, he would not hesitate at all.

The manner of performing examples will appear new to many, but
it will be found much more agreeable to the practice of men of busi-

ness, and men of science generally, than those commonly found in books. This is the method of those that understand the subject. The others were invented as a substitute for understanding.

The *rule of three* is entirely omitted. This has been considered useless in France, for some years, though it has been retained in their books. Those who understand the principles sufficiently to comprehend the nature of the rule of three, can do much better without it than with it, for when it is used, it obscures, rather than illustrates, the subject to which it is applied. The principle of the rule of three is similar to the combinations in Art. XVI.

The rule of Position has been omitted. This is an artificial rule, the principle of which cannot be well understood without the aid of Algebra: and when Algebra is understood, Position is useless. Besides, all the examples which can be performed by Position, may be performed much more easily, and in a manner perfectly intelligible, without it. The manner in which they are performed is similar to that of Algebra, but without Algebraic notation. The principle of *false position*, properly so called, is applied only to questions where there are not sufficient data to solve them directly.

Powers and roots, though arithmetical operations, come more properly within the province of Algebra.

There are no answers to the examples given in the book. A key is published separately for teachers, containing the answers and solutions of the most difficult examples.

TABLE OF CONTENTS.

(This Table equally refers to Parts I. and II.)

INDEX TO PARTICULAR SUBJECTS.

—

ARITHMETIC.

PART I.

THE student may perform the following examples in his mind.

1. James has 3 cents and Charles has 5; how many have they both?

2. Charles bought 3 bunns for 16 cents, a quart of cherries for 8 cents, and 2 oranges for 12 cents; how many cents did he lay out?

3. A man bought a hat for 8 dollars, a coat for 27 dollars, a pair of boots for 5 dollars, and a vest for 7 dollars; how many dollars did the whole come to?

4. A man bought a firkin of butter for 8 dollars, a quarter of veal for 45 cents, and a barrel of cider for 3 dollars and 25 cents; how much did he give for the whole?

5. A man sold a horse for 127 dollars, a load of hay for 15 dollars, and 3 barrels of cider for 12 dollars; how much did he receive for the whole?

6. A man travelled 27 miles in one day, 15 miles the next day, and 8 miles the next; how many miles did he travel in the whole?

7. A man received 42 dollars and 37 cents of one person, 4 dollars and 68 cents of another, and 7 dollars and 83 cents of a third; how much did he receive in the whole?

8. I received 25 dollars and 58 cents of one man, 45 dollars and 83 cents of another, and 8 dollars and 39 cents of a third; how much did I receive in the whole?

The two last examples may be performed in the mind, but they will be rather difficult. A more convenient method will soon be found.

NUMERATION.

1. Write in words the following numbers.

1	27	24	10,000
2	35	25	20,030
3	58	26	50,705
4	63	27	67,083
5	70	28	300,050
6	84	29	476,089
7	96	30	707,720
8	100	31	1,000,370
9	103	32	5,600,073
10	110	33	8,081,305
11	113	34	59,006,341
12	127	35	305,870,400
13	308	36	590,047,608
14	520	37	1,000,000,000
15	738	38	3,670,000,387
16	1,000	39	45,007,070,007
17	1,001	40	680,930,100,700
18	1,010	41	50,787,657,000,500
19	1,100	42	270,000,838,003,908
20	1,018	43	68,907,605
21	2,107	44	58,000,034,750
22	3 250	45	6,703,720,000,857
23	5,796		

Write in figures the following numbers.

1. Thirty-four.
2. Fifty-seven.
3. Sixty-three.
4. Eighty.
5. One hundred.
6. One hundred and one.
7. One hundred and ten.
8. Three hundred and eleven.
9. Five hundred and seventeen.
10. Eight hundred and fifty.
11. Nine hundred and eighty-six.
12. One thousand and one.
13. One thousand and ten.
14. Three thousand, one hundred and one.
15. Five thousand and sixty.

16. Ten thousand and five.

17. Thirty thousand, five hundred, and four.

18. Sixty-seven thousand, and forty.

19. Five hundred thousand, and seventy-one.

20. Two hundred and seven thousand, six hundred.

21. Four millions, sixty thousand, and eighty-four.

22. Ninety-seven millions, thirty-five thousand, eight hundred and five.

23. Fifty millions, seventy thousand, and eight.

24. Three hundred millions, and fifty-seven.

25. Two billions, fifty-three millions, three hundred and five thousand, two hundred.

26. Fifty billions, two hundred and seven millions, sixty-seven thousand, two hundred.

27. Eighty-seven millions, and sixty-three.

28. Six hundred billions, two hundred and seven thousand, and three.

29. Thirty-five trillions, nine millions, and fifty-eight.

30. Six hundred and fifty-seven trillions, seven billions, ninety-seven thousand, and sixty-seven.

31. Seventy millions, two hundred and fifty thousand, three hundred and sixty-seven.

32. Four hundred and seven trillions, and eighty-seven thousand.

33. Thirty-five billions, ninety-eight thousand, one hundred.

34. Forty millions, two hundred thousand, and seventy-four.

35. Eighty-three millions, seven hundred and sixty-three thousand, nine hundred and fifty-seven.

ADDITION.

II. 1.* A man bought a watch for fifty-eight dollars, a cane for five dollars, a hat for ten dollars, and a pair of boots for six dollars. What did he give for the whole?

2. In an orchard there are six rows of trees; in the two first rows, there are fifteen trees in each row; in the third row, seventeen; in the fourth row, eleven; in the fifth row,

* See First Lessons. sect. I.

eight; and in the sixth row, nineteen. How many trees are there in the orchard?

3. Four men bought a piece of land; the first gave sixty-three dollars; the second, seventy-eight; the third, forty-five; and the fourth, twenty-three. How much did they give for the land?

4. In an orchard, 19 trees bear cherries, twenty-eight bear peaches, 8 bear plums, and 54 bear apples. How many trees are there in the orchard?

5. How many days are there in a year, there being in January 31 days; in February 28; in March 31; in April 30; in May 31; in June 30; in July 31; in August 31; in September 30; in October 31; in November 30; in December 31?

6. The distance from Portland (in Maine) to Boston, is 114 miles; from Boston to Providence, 40 miles; from Providence to New Haven 122 miles; from New Haven to New York, 88 miles; from New York to Philadelphia, 95 miles; from Philadelphia to Baltimore, 102 miles; from Baltimore to Charleston, S. C. 716 miles; from Charleston to Savannah, 110 miles. How many miles is it from Portland to Savannah?

7. What number of dollars are there in four bags; the first containing 275 dollars; the second, 356; the third, 178; the fourth, 69?

8. How many times does the hammer of a clock strike in 24 hours?

Note. At 1 o'clock it strikes once, at 2 o'clock it strikes twice, &c.

9. A man has four horses; the first is worth sixty-seven dollars; the second is worth eighty-four dollars; the third is worth one hundred and twenty dollars; and the fourth is worth one hundred and eighty-seven dollars; and he has four saddles worth twelve dollars apiece. How much are the horses and saddles worth?

10. A man owns five houses; for the first he receives a rent of 427 dollars; for the second, 763 dollars; for the third, 654 dollars; for the fourth, 500 dollars; and for the fifth, 325 dollars; and the rest of his income is 3,250 dollars. What is his whole income?

11. A gentleman owns five farms; the first is worth 11,500 dollars; the second, 3,057 dollars; the third, 2,468 dollars; the fourth, 9,462 dollars; and the fifth, 850 dollars;

and he owns a house worth 15,000 dollars, a carriage worth 753 dollars, and two horses worth 175 dollars apiece. How much are they all worth ?

12. A merchant bought four pieces of cloth, each piece containing 57 yards. For the first piece he gave 235 dollars; for the second, 384 dollars; for the third, 327 dollars; and for the fourth, 486 dollars. How many yards of cloth did he buy ? How much did he give for the whole ?

13. In 1818 the navy of the United States consisted of three 74s; five 44 gun frigates; three 36s; two 32s; one 20; ten 18s. How many guns did they all carry ?

14. Suppose it requires 650 men to man a 74; 475 to man a 44; 275 to man a 36; 350 to man a 32; 200 to man a 20; and 180 to man an 18. How many men would it take to man the whole ?

15. The hind quarters of a cow weighed one hundred and five pounds each ; the fore quarters weighed ninety-four pounds each; the hide weighed sixty-three pounds ; and the tallow seventy-six pounds. What was the whole weight of the cow ?

16. A man bought a barrel of flour for 6 dollars, and sold it so as to gain 2 dollars. How much did he sell it for ?

17. I bought a quantity of salt for 18 dollars, and sold it for 7 dollars more than I gave for it; how much did I sell it for ?

18. A man bought three hogsheads of molasses for 132 dollars, and sold it so as to gain 25 dollars; how much did he sell it for ?

19. A man being asked his age, answered that he was twenty-seven years old when he was married, and that he had been married thirty-nine years. How old was he ?

20. A man being asked his age, answered that he had passed the 19 first years of his life in America, and that he had afterwards spent 7 years in Germany, 13 years in France, 3 years in Holland, and 24 years in England. How old was he ?

21. A merchant bought four hogsheads of wine for four hundred and thirty-seven dollars, and sold it again for ninety-four dollars more than he gave for it. How much did he sell it for ?

22. A man commenced trade with three thousand, two hundred and fifty dollars ; after trading for some time, he

found he had gained two hundred and thirty-seven dollars.
How much had he then ?

23. Money was first made at Argos, eight hundred and
ninety-four years before Christ. . How long has it been in
use at this date, 1822 ?

24. The war between Great Britain and the American
colonies commenced in 1775 and continued 8 years. In
what year was the war concluded ?

25. General Washington was born in the year 1732, and
was 67 years old when he died. In what year did he die ?

26. The first tragedy was acted at Athens, on a cart, by
Thespis, five hundred and thirty-six years before Christ.
How many years is it since ?

27. What was the number of inhabitants in the New
England States, in 1820, there being in

Maine	298,335
New Hampshire	244,161
Vermont	235,764
Massachusetts	523,287
Rhode Island	83,059
Connecticut	275,248 ?

28. What was the number of inhabitants in the Middle States, there being in

New York	1,372,812
New Jersey	277,575
Pennsylvania	1,049,398
Delaware	72,749
Maryland .	407,350 ?

29. What was the number of inhabitants in the following
States, there being in

Virginia	1,065,366
North Carolina	638,829
South Carolina	490,309
Georgia	340,989
Kentucky	564,317
Tennessee	422,813
Alabama	127,901
Mississippi	75,448
Louisiana	153,407 ?

30. What was the number of inhabitants in the following
States, there being in

Ohio	581,434
Indiana	147,178

Illinois	55,211
Missouri	66,586
Arkansas Territory	14,273
Michigan Territory	8,896
District of Columbia	33,039 ?

31. What was the whole number of inhabitants in the United States in 1820 ?

32. Add together the following numbers ; 32,753 ; 2,047 ; 840,397 ; 47,640.

33. What is the sum of the following numbers ; 30 ; 843 ; 30,804 ; 387,643 ; 13 ; 8,406,127 ; 4 ; 900,600 ?

34. What is the sum of the following numbers, three millions and seven thousand ; thirty-five ; four hundred and eighty-seven ; two thousand and forty-three ; ninety-six millions, thirty-four thousand, and forty-two ; and seventeen ?

———————

MULTIPLICATION.

III.   1.* What will two barrels of rum cost, at 27 dollars a barrel ?

2. What will 3 hogsheads of molasses amount to, at 26 dollars a hogshead ?

3. What will 14 pounds of veal come to, at 4 cents a pound ?

4. What will seventeen pounds of beef cost, at five cents a pound ?

5. What will five cows cost at 19 dollars apiece ?

6. What will 3 oxen cost at 47 dollars apiece ?

7. What cost 15 yards of cloth at 8 dollars a yard ?

8. What cost 26 barrels of cider at 4 dollars a barrel ?

9. What cost 98 barrels of flour at 7 dollars a barrel ?

10. What cost 794 barrels of flour at 9 dollars a barrel ?

11. There is an orchard consisting of 9 rows of trees, and there are 57 trees in each row. How many trees are there in the orchard.

12. A man bought 8 pieces of cloth, each piece containing 38 yards, at 7 dollars a yard. How many yards were there, and what did he give for the whole ?

13. A man bought 9 pieces of broadcloth, each piece con-

* See First Lessons, sect. II.

2 *

taining 47 yards, at 6 dollars a yard ; and 25 barrels of flour
at 7 dollars a barrel.   What did he give for the whole ?

14. A merchant bought a hogshead of wine, at the rate of
2 dollars a gallon ; what did it come to ?

WINE MEASURE.

4	gills (gl.)	make	1 pint marked	pt.
2	pints		1 quart	qt.
4	quarts		1 gallon	gal.
31½	gallons		1 barrel or half hhd.	bbl.
63	gallons		1 hogshead	hhd.
2	hogsheads		1 pipe or butt	p. or b.
2	pipes		1 tun	T.

By this measure brandy, spirits, perry, cider, mead, vinegar, and oil are measured.

15. At 3 dollars a gallon, what will 2 pipes of wine cost?

16. At 4 cents a gill, what will 1 pint of brandy cost?

17. At 5 cents a gill what will 1 quart of wine cost?
What will 1 gallon cost?

Note.   Since 100 cents make 1 dollar, it will be easy to
tell how many dollars there are in any number of cents.

18. At 8 cents a quart, what will 1 hhd. of molasses
come to ?

19. How many pints are there in 87 quarts?

20. How many gills are there in 174 pints?

21. How many quarts are there in 1 hhd. of wine ?

22. How many quarts are there in 4 hhds. of brandy ?

23. How many pints are there in one hhd. of molasses ?

24. How many pints are there in 1 pipe ?

25. How many gills are there in 1 hhd ?

26. How many gills are there in 1 T. ?

27. How many quarts in 8 gals. 2 quarts ?

28. How many pints in 4 gals. 3 qts. 1 pint ?

29. How many gallons in 3 hhds. 42 gals. ?

30. How many quarts in 1 p. 1 hhd. ?

31. How many pints in 1 hhd. 35 gals. 3 qts. 1 pt. ?

32. How many gills in 3 hhds. 27 gals. 1 qt. 1 pt. 3 gls. ?

33. A man having 1 T. of wine, retailed it at 5 cents a
gill, how much did it come to ?

34. A man bought a quarter of beef, weighing 237 pounds,
at 7 cents a pound.   How much did it cost ?

35. How many are 3 times 784 ?

36. How many are 5 times 1,328?
37. How many are nine times 87,436?
38. Multiply 2,487 by 8.
39. Multiply 820,438 by 7.
40. Multiply 13,052,068 by 5.

IV.　1. What will 18 oxen cost at 57 dollars apiece?
Note. Find first what 6 oxen will cost, and 18 oxen will cost 3 times as much. Perform the following examples in a similar manner.

2. What would 14 chests of tea cost, at 87 dollars a chest?

3. A merchant bought 84 pieces of linen, at 16 dollars apiece; how much did it come to?

4. A merchant bought 15 hogsheads of wine, at 97 dollars a hogshead. How much did the whole amount to?

5. A merchant sold 20 hhds. of brandy at 2 dollars a gallon. How much did each hogshead amount to? How much did the whole amount to?

6. What would 28 bales of cotton come to, at 75 dollars a bale?

TIME.

60 seconds (sec.) make	1 minute,	marked	min.
60 minutes	1 hour		h.
24 hours	1 day		d.
7 days	1 week		w.
4 weeks	1 month		mo.
13 months 1 day & 6 hours, or 365 days and 6 hours }	1 year		y.
12 calendar months	1 year		

7. If a man can earn eight dollars in a week, how much can he earn in a year?

8. If the expenses of a man's family are 32 dollars a week, what will they amount to in a year? What in 2 years?

9. How many hours are there in a week?

10. How many minutes are there in a day?

11. How many minutes are there in a week?

12. How many hours are there in 2 mo. 3 d.?

13. If a man can travel 7 miles in an hour, how far can he travel in 8 days, when the days are 9 hours long?

14. If a ship sail 11 miles in an hour, how far would it sail at that rate in one day, or 24 hours?

15. If a ship sail 8 miles in an hour, how far would it sail at that rate in 18 days?

16. Suppose a cistern has a cock which conveys 37 gallons into it in an half hour, how much would run into it in 1 d. 8 h.

17. If a man can earn 18 dollars in a calendar month, how much would he earn in 7 y. 8 mo. ?

18. In one year how many minutes ?

19. In two y. 3 mo. 18 d. how many days ?

20. A cannon ball at its first discharge, flies at the rate of about 9 miles in a minute; how far would it fly at that rate in 24 hours ? How far in 15 days ?

21.	Multiply	87 by 14	32.	Multiply	21,378 by 36
22.		321 15	33.		825 42
23.		463 16	34.		164 45
24.		275 18	35.		1,163 48
25.		144 21	36.		9,876 49
26.		2,107 24	37.		40,073 54
27.		381 25	38.		3,502 56
28.		1,234 27	39.		127 63
29.		3,002 28	40.		308 72
30.		4,381 32	41.		1,437 81
31.		11,962 35			

42. What would 17 loads of hay come to at 26 dollars a load ?

Note. First find the price of 16 loads, and then add the price of 1 load. Perform the following examples in a similar manner.

43. What would 17 oxen cost, at 87 dollars apiece ?

44. What would 87 pounds of tobacco cost, at 23 cents a pound ?

45. What would 28 pounds of sugar cost, at 13 cents a pound ?

46. What would 59 bushels of potatoes cost, at 38 cents a bushel ?

47. What costs 1 hhd. of molasses at 37 cents a gallon ?

48.	Multiply	19 by 17	52.	Multiply	206 by 38
49.		37 19	53.		314 47
50.		106 23	54.		203 58
51.		141 34	55.		715 67

V. 1. What cost 5 pounds of beef at 10 cents a pound ?

2. What will 12 barrels of flour come to, at 10 dollars a *barrel ?*

Note. Observe that when you multiply by 10, it is done by annexing a zero to the right of the number ; and by 100, it is done by annexing two zeros, &c. ; and find the reason why.

3. What would a hogshead of wine come to, at ten cents a pint ?

4. If 10 men can do a piece of work in 7 days, how many days will it take 1 man to do it ?

5. What would an ox, weighing 873 pounds, come to, at 10 cents a pound ?

6. If 100 men were to receive 8 dollars apiece, how many dollars would they all receive ?

7. If 27 men were to receive 100 dollars apiece, how many dollars would they all receive ?

FEDERAL MONEY.

10 mills (m.)	make	1 cent	marked	c.
10 cents		1 dime		d.
10 dimes		1 dollar		dol. or $.
10 dollars		1 eagle		E.

8. In 3 dimes how many cents ?

9. In 5 dollars how many dimes ? How many cents ?

10. In 17 dollars how many cents ?

11. In 83 cents how many mills ?

12. In 753 dols. how many cents ?

13. In 1 dol. how many mills ?

14. In 84 dols. how many mills ?

15. In 7 dols. and 53 cents, how many cents ?

16. In 183 dols. and 14 cents, how many cents ?

17. In 283 dols. 43 cents and 8 mills, how many mills ?

18. In 8246 dols. 2 d. 5 c. 6 m. how many mills ?

It is usual to write dollars and cents in the following manner : 43 dols. 5. d. 7 c. and 4 mills, is written $43.574. The character $ written before shows that it is federal money. The figures at the left of the point (.) are so many dollars, the first figure at the right of the point is so many dimes, the next so many cents, and the third so many mills.

It may be observed that when dollars stand alone, they are changed to dimes by annexing one zero to the right, because that multiplies them by 10. They are changed to *cents by annexing two zeros, because that multiplies them*

by 100. They are changed to mills by annexing three zeros, because that multiplies them by 1,000. Thus 43 dollars are 430 dimes, 4,300 cents, or 43,000 mills. 5 dimes
are 50 cents, or 500 mills. 7 cents are 70 mills. The
above example then may be read 43 dols. 57 cents and 4
mills; or 435 dimes, 7 cents, and 4 mills; or 4,357 cents
and 4 mills; or 43,574 mills. When there are dollars,
dimes, and cents, the figures on the left of the point may be
read dollars, and those on the right, cents; or they may be
all read together as cents. When the number of cents exceeds 100, they are changed to dollars by putting a point
between the second and third figures from the right. If,
there are mills in the number, all the figures may be read
together as mills. Any number of mills are changed to dollars by putting a point between the third and fourth figure
from the right; the figures at the left will be dollars, and
those at the right, dimes, cents, and mills. Since any sum
which has cents or mills in it, may be considered as so many
cents or mills, it is evident that any operation, as addition,
multiplication, &c. may be performed upon it in the same
manner as upon simple numbers.

If the sum consists of dollars and a number of cents less
than ten, there must be a zero between the dollars and the
cents in the place of dimes. Thus 7 dols. and 5 cents must
be written $7.05.

19. What will 10 yards of cloth cost at $4.53 a yard?
20. What will 10 pounds of coffee cost at $0.27 a pound?
21. What will 100 sheep cost at $8.45 apiece?
22. What will 1,000 yards of cloth cost at $0.35 a yard?

23.	Multiply	5	by 10	32.	Multiply	90	·by	100
24.		47	10	33.		4		1,000
25.		30	10	34.·		73		1,000
26.		124	10	35.		80	.	1,000
27.		387	10	36.		- 132		1,000
28.		450	10	37.		800		1,000
29.		13,008	10	38.		1,643		1,000
30.		7	100	39.		725		10,000
31.		38	100	40.		76,438		10,000

VI. 1. What cost 75 lb. of tobacco at 20 cents a pound?
2. What cost 30 cords of wood at $6,75 a cord?
3. If 400 men receive 135 dollars apiece, how many dollars will they all receive?

4. If 30 men can do a piece of work in 43 days, how many days will it take 1 man to do it?

5. If 70 men can do a piece of work in 83 days, how many men will it take to do it in one day?

6. If the pendulum of a clock swing once in a second, how many times will it swing in an hour? How many times in a day? How many times in a week?

7. How many seconds are there in 10 min. 23 sec. ?

8. How many minutes are there in 7 h. 23 min. ?

9. How many minutes are there in 3d. 7 h. 43 min. ?

10. How many seconds are there in 8 d. 7 h. 34 min. 19 sec. ?

11. A garrison of 3,000 men are to be paid, and each man is to receive 128 dollars. How many dollars will they a'll receive?

12. What cost 30 barrels of cider at $3.50 a barrel?

13. There are 320 rods in a mile, how many rods are there in 7 miles? How many in 10 miles? How many in 30 miles? How many in 500 miles?

14. Multiply	34	by	20	18. Multiply	4,007	by	80
15.	57		300	19.	11,600		700
16.	250		60	20.	4,960		40,000
17.	387		5,000	21.	13,400		8,000

VII.   1. What will 17 oxen come to at 42 dollars apiece?

Note. Find the price of 10 oxen and of 7 oxen separately, and then add them together.

2. What will 34 barrels of flour come to, at $6.43 a barrel?

Note. Find the price of 30 barrels and of 4 barrels separately, and then add them together.

3. What cost 19 gallons of wine, at $1.28 a gallon?

4. What cost 68 yards of cloth, at $9.36 a yard?

5. What will 87 thousand of boards come to, at $5.50 a thousand?

6. What will 58 barrels of beef come to, at $9.75 a barrel?

7. What will 87 gallons of brandy come to, at $1.60 a gallon?

8. A and B depart from the same place and travel in opposite directions, A at the rate of 38 miles in a day, and B at the rate of 42 miles a day. How far apart will they be at the end of the first day? How far at the end of 15 days?

9. What will 287 barrels of turpentine come to, at $3.25 a barrel ?

Note. Find the price of 200 barrels, of 80 barrels, and of 7 barrels separately, and then add them together.

10. What will 358 barrels of beef come to, at $7.55 a barrel ?

11. A drover bought 853 sheep at an average price of $3.58 apiece. What were the whole worth?

12. A merchant bought 105 hundred weight of lead, at $17.33 a hundred weight ; how much did the whole come to ?

13. If a ship sail 8 miles in an hour, how many miles will she sail in a day, at that rate ? How far in 127 days ?

14. An army of 8,975 men are to receive 138 dollars apiece. How many dollars will they all receive ?

15. An army of 11,327 men are to receive a year's pay, at the rate of 5 dollars a month for each man. How many dollars will they all receive ?

16. Bought 207 chaldrons of coal, at $12.375 a chaldron. How much did it come to ?

17. Bought 857 pounds of sugar at $0.125 a pound. How much did it come to ?

18. Shipped 350 casks of butter worth $14.50 a cask. What was the value of the whole ?

19. What cost 354 fother of lead, at $63.57 a fother ?

20. What cost 25,837 gallons of brandy, at $2.375 a gallon ?

21. If it cost $28.56 to clothe a soldier 1 year, how many dollars will it cost to clothe an army of 15,200 men the same time ?

22.	Multiply	887	by	47
23.		6,300		250
24.		1,006		398
25.		15,030		1,005
26.		38,446		2,700
27.		487,500		38,400
28.		7,035,064		30,704
29.		9,800,000		37,000
30.		78,508,060		300,005
31.		43,060,085		703,004

Miscellaneous Examples.

1. If 1 pound of tobacco cost 28 cents, what will a keg of tobacco, weighing 112 pounds, cost ?

AVOIRDUPOIS WEIGHT.

16 drams (dr.) make	1 ounce,	marked	oz.
16 ounces	1 pound		lb.
28 pounds	1 quarter		qr.
4 quarters	1 hundred weight		cwt.
20 hundred weight	1 ton		T.

By this weight are weighed all things of a coarse and drossy nature; such as butter, cheese, flesh, grocery wares, and all metals except gold and silver.

2. At 12 cents per lb. how much will 1 quarter of sugar come to?

3. If 1 quarter of sugar cost 7 dollars, what will 1 cwt cost?

4. How many pounds are there in 1 cwt.?

5. In 2 cwt. 2 qrs. how many quarters?

6. In 3 qrs. 18 lb. how many pounds?

7. In 2 cwt. 1 qr. how many pounds?

8. In 1 cwt. 3 qrs. 23 lb. how many pounds?

9. In 18 lb. how many ounces?

10. In 12 cwt. how many ounces?

11. In 14 cwt. 3 qrs. 15 lb. 8 oz. how many ounces?

12. At 9 cents a pound, what cost 3 cwt. 2 qrs. 16 lb. of sugar?

TROY WEIGHT.

24 grains (gr.) make	1 penny-weight, marked	dwt.
20 penny-weights	1 ounce	oz.
12 ounces	1 pound	lb.

By this weight are weighed gold, silver, jewels, corn, bread, and liquors.

13. If an ingot of silver weigh 42 oz. 13 dwt., what is it worth at 4 cents per dwt.?

14. What is the value of a silver cup weighing 9 oz. 4 dwt. 16 gr. at 3 mills per grain?

15. In 15 ingots of gold each weighing 9 oz. 5 dwt. 7 gr. how many grains?

APOTHECARIES' WEIGHT.

20 grains (gr.) make	1 scruple, marked	sc.
3 scruples	1 dram	dr. or ℈
8 drams	1 ounce	oz. or ℥
12 ounces	1 pound	℔.

3

Apothecaries use this weight in compounding their medicines, but they buy and sell by Avoirdupois weight. Apothecaries' is the same as Troy weight, having only some different divisions.

16. In 9℔. 8 ℨ. 1 ℨ. 2 sc. 19 gr. how many grains?

DRY MEASURE.

2 pints (pt.)	make	1 quart, marked	qt.
8 quarts		1 peck	pk.
4 pecks		1 bushel	bu.
8 bushels		1 quarter	qr.

By this measure, salt, ore, oysters, corn, and other dry goods are measured.

17. At 43 cents a peck, what cost 14 bu. 3 pks. of wheat?
18. At 3 cents a quart what will 5 bu. 2 pks. 3 qts. of salt come to?

CLOTH MEASURE.

2¼ inches (in.)	make	1 nail, marked	nl.
4 nails		1 quarter	qr.
4 quarters		1 yard	yd.
3 quarters		1 ell Flemish	Ell Fl.
5 quarter:		1 ell English	Ell Eng.
5 quarter		1 aune or ell French.	

19. At 27 cents a nail, what is the price of 2 yds. 1 qr. 3 nls. of cloth.
20. I. 1 qr. cost $2,50, what cost 43 ells English of broadcloth?
21. At 42 cents a nail, what cost 13 ells Fl. 3 qrs. of broadcloth?
22. How many seconds are there in 4 years?
23. How many seconds are there in 8 y. 3 mo. 2 wks. 2 d. 19 h. 43 min. 57 sec.?
24. How many calendar months are there from the 1st Feb. 1819, to the 1st August, 1822?
25. How many days are there from the 7th Sept. 1817, to the 17th May, 1822?
26. How many minutes are there from the 13th July, at 43 minutes after 9 in the morning, to the 5th Nov. at 19 min. past 3 in the afternoon?

27. How many seconds old are you ?

28. How many seconds from the commencement of the Christian era to the year 1822 ?

29. At 4 cents an ounce, how much would 3 cwt. 2 qrs. 18 lb. 7 oz. of snuff come to ?

30. At 28 cents a pound, what would 3 T. 2 cwt. 3 qrs. 16 lb. of tobacco come to ?

31. If a cannon ball flies 8 miles in a minute, how far would it fly at that rate in 7 y. 2 mo. 3 wks. 2 days ?

32. If a quantity of provision will last 324 men 7 days, how many men will it last one day ?

33. A garrison of 527 men have provision sufficient to last 47 days, if each man is allowed 15 oz. a day ; how many days would it last if each man were allowed only 1 oz. a day ?

34. A garrison of 527 men have provision sufficient to last 47 days, if each man is allowed 15 oz. a day ; how many men would it serve the same time, if each man were allowed only 1 oz. a day ?

35. If a man performs a journey in 58 days, by travelling 9 hours in a day, how many hours is he performing it ?

36. If by working 13 hours in a day a man can perform a piece of work in 217 days ; how long would it take him to do it if he worked only 1 hour in a day ?

37. If by labouring 14 hours in a day 237 men can build a ship in 132 days, how many days would it take them, if they work only 1 hour in a day ?   How many men would it take to do it in 132 days, if they work only 1 hour in a day ?

38. How many yards of cloth that is 1 qr. wide, are equal to 27 yards that is 1 yd. wide ?

39. If a piece of cloth that is 1 qr. wide is worth $67.25, what is a piece containing the same number of yards of the same kind of cloth worth, that is 1 yd. wide ?

40. If a bushel of wheat afford 65 eight-penny loaves, how many penny loaves may be obtained from it ?

41. What is the price of 4 pieces of cloth, the first containing 21 yards, at $4.75 a yard ; the second containing 27 yards, at $7.25 a yard ; the third containing 18 yards, at $9.00 a yard ; and the fourth containing 32 yards, at $8.57 a yard ?

42. A man bought 15 lb. of beef, at 9 cents a pound ; 28 lb. of sugar, at $0.125 a pound ; 18 gallons of wine, at

$1.56 a gallon; a barrel of flour, for $8.00; and 3 barrels of cider, at $3.50 a barrel. How much did the whole amount to?

Interest is a reward allowed by a debtor to a creditor for the use of money. It is reckoned by the hundred, hence the rate is called so much *per cent.* or *per centum.* *Per centum* is Latin, signifying by the hundred. 6 per cent. signifies 6 dollars on a hundred dollars, 6 cents on a hundred cents, £6 on £100, &c. so 5 per cent. signifies 5 dollars on 100 dollars, &c. *Insurance, commission,* and *premiums* of every kind are reckoned in this way. *Discount* is so much per cent. to be taken out of the principal.

43. If 1 dollar gain 6 cents interest a year, how much will 13 dollars gain in the same time?

44. What is the interest of $43.00 for 1 year at 6 per cent.?

45. What is the interest of $157.00 for 1 year at 5 per cent.?

46. What is the interest of $1.00 for 2 years at 6 per cent.? What for 5 years?

47. What is the interest of $247.00 for 3 years at 7 per cent?

48. How much must I give for insuring a ship and cargo worth $150,000.00 at 2 per cent.?

49. Imported some books from England, for which I paid $150.00 there. The duties in Boston were 15 per cent., the freight $5.00. What did the books cost me?

50. What must I receive for a note of $275.00 that has been due 3 years; interest at 6 per cent.?

51. A man failing in trade, is able to pay only $0.68 on a dollar; how much can he pay on a debt of $5 dollars? How much on a debt of 20 dollars?

52. A man failing in trade, is able to pay only $0.73 on a dollar; how much will he pay on a debt of $47.00? How much on a debt of $123.00? How much on a debt of $2,500.00?

53. A merchant bought a quantity of goods for 243 dollars, and sold them so as to gain 15 per cent.; how much did he gain, and how much did he sell them for?

54. A merchant bought a quantity of goods for $843.00; how much must he sell them for to gain 23 per cent.?

VIII. 1.* David had nine peaches, and gave four of them to George ; how many had he left ?

2. A man having 15 dollars, lost 9 of them ; how many had he left ?

3. David and William counted their apples; David had 35, and William had 17 less ; how many had William ?

4. A man owing 48 dollars, paid 29 ; how many did he then owe ?

5. A man owing 48 dollars, paid all but 19 ; how many did he pay ?

6. A man owing a sum of money, paid 29 dollars, and then he owed 19 ; how many did he owe at first ?

7. A man being asked how old he was when he was married, answered, that his present age was sixty-four years, and that he had been married 37 years ; what was his age when he was married ?

8. A man being asked how long he had been married, answered, that his present age was sixty-four years, and that he was twenty-seven years old when he was married ; how long had he been married ?

9. A man being asked his age, answered, that he was 27 years old when he was married, and that he had been married 37 years. What was his age ?

10. A man bought a piece of cloth containing 93 yards, and sold 45 yards of it ; how many yards had he left ?

11. A merchant bought a piece of cloth for one hundred and fifteen dollars, and sold it again for one hundred and thirty-eight dollars. How much did he gain by the bargain ?

12. A merchant sold a piece of cloth for 138 dollars, which was 23 dollars more than he gave for it; how much did he give for it ?

13. A merchant bought a piece of cloth for 115 dollars, and sold it so as to lose 23 dollars. How much did he sell it for ?

14. A man bought a quantity of wine for 753 dollars, but not being so good as he expected, he was willing to lose 87 dollars in the sale of it ; how much did he sell it for ?

15. A man owing two thousand, six hundred, and forty-

* See First Lessons, sect. 1.

three dollars, paid at several times as follows; at one time two hundred and seventy-five dollars; at another fifty-eight dollars; at another seven dollars; and at another one thousand and sixty-seven dollars; how much did he then owe?

16. From Boston to Providence it is 41 miles, and from Boston to Attleborough (which is upon the road from Boston to Providence) it is 28 miles; how   is it from Attleborough to Providence?

17. From Boston to New York it is   0 miles; suppose a man to have set out from Boston for New York, and to have travelled 14 hours, at the rate of five miles in an hour; how much farther has he to travel?

18. General Washington was born A. D. 1732, and died in 1799; how old was he when he died?

19. Dr. Franklin died A. D. 1790, and was 84 years old when he died; in what year was he born?

20. A gentleman gave 853 dollars for a carriage and two horses; the carriage alone was valued at 387 dollars; what was the value of the horses? How much more were the horses worth than the carriage?

21. A man died leaving an estate of eight thousand, four hundred, and twenty-three dollars; which he bequeathed as follows; two thousand, three hundred dollars to each of his two daughters, and the rest to his son; what was the son's share?

22. A gentleman bought a house for sixteen thousand, and twenty-eight dollars; a carriage for three hundred and eight dollars, and a span of horses for five hundred and eighty-three dollars. He paid as follows; at one time ninety-seven dollars; at another, one thousand, and eight dollars; and at a third, four thousand, two hundred, and six dollars. How much did he then owe?

23. In Boston, by the census of 1820, there were 43,278 inhabitants; in New York, 123,706. How many more inhabitants were there in New York than in Boston?

24. In Boston, by the census of 1810, the number of inhabitants was 33,250; and in 1820 it was 43,278. What was the increase for 10 years?

25. A merchant bought 2 pipes of brandy for 642 dollars, and retailed it at 3 dollars a gallon. How much did he gain?

26. A man bought 359 kegs of tobacco, at 9 dollars a keg; 654 barrels of beef, at 8 dollars a barrel; 9 bags of coffee, at 29 dollars a bag. In exchange he gave 3 hhds.

, of brandy, at 2 dollars a gallon ; 473 cwt. of sugar, at 8 dollars per cwt. How much did he then owe ?

27. A man bought 7 lb. of sugar, at $0.125 per lb. ; 4 gals. of molasses, at 0.375 per gall. ; 5 lb. of raisins, at $0.14 per lb. ; 1 bbl. of flour, for $6.00. He paid a ten dollar bill ; how much change ought he to receive back ?

28. Two merchants, A and B, traded as follows ; A sold B 24 pipes of wine, at $1.87 per gal. ; and B sold A 32 hhds.' of molasses, at $47.00 per hhd. The balance was paid in money ; how much money was paid, and which received it ?

29. A merchant sold 35 barrels of flour, at 7 dollars per barrel ; but for ready money he made 10 per cent. discount. How much did the flour come to after the discount was made ?

30. A merchant bought 15 hhds. of wine, at $2.00 per gallon ; but not finding so ready a sale as he wished, he was obliged to sell it so as to lose 8 per cent. on the cost. How much did he lose, and how much did he sell the whole for ?

31. Suppose a gentleman's income is $1,836.00 a year, and he spends $3.27 a day, one day with another ; how much will he spend in the year ? How much of his income will he save ?

32. What is the difference between 487,068 and 24,703 ?

33. How much larger is 380,064 than 87,065 ?

34. How much smaller is 8.756 than 3",005,078 ?

35. How much must you add to 7,643 to make 16,487 ?

36. How much must you subtract from 2,483 to leave 527 ?

37. If you divide 3,880 dollars between two men, giving one 1,907 dollars, how much will you give the other ?

38. Subtract 38,506 from 90,000.

39. Subtract 20,076 from 180,003.

40. A man having 1,000 dollars, gave away one dollar ; how many dollars had he left ?

41. A man having $1,000.00, lost seventeen cents, how much had he left ?

42. What is the difference between 13 and 800,060 ?

43. What is the difference between 160,000 and 70 ?

44. How much must you add to 123 to make 10,000 ?

45. A man's income is $2,738.43 a year, and he spends $1,897.57 ; how much does he save ?

46. Subtract 93 from 80,640.

47. A merchant shipped molasses to the amount of $15,000.00, but during a storm the master was obliged to throw overboard to the amount of $853.42; what was the value of the remaining part?

48. A man bought goods to the amount of $1,153.00, at 6 months' credit, but preferring to pay ready money, a dis.. count was made of $35.47. What did he pay for the goods?

49. Subtract one cent from a thousand dollars.

<hr>

DIVISION.

IX.   1. How many oranges, at 6 cents apiece, can you buy for 36 cents?

2. How many barrels of cider, at 3 dollars a barrel, can be bought for 27 dollars?

3. How many bushels of apples, at 4 shillings a bushel, can you buy for 56 shillings?

4. How many barrels of flour, at 7 dollars a barrel, can you buy for 98 dollars?

5. How many dollars are there in 96 shillings?

ENGLISH MONEY.

4 farthings (qr.)	make	1 penny,	marked d.
12 pence		1 shilling	s.
20 shillings		1 pound	£
21 shillings		1 guinea.	

This money was used in this country until A. D. 1786, when, by an act of Congress, the present system, which is called *Federal Money*, was adopted. Some of these denominations, however, are still used in this country, as the shilling and the penny, but they are different in value from the English. In English money 4s. 6d. is equal in value to the Spanish and American dollar. But a dollar is called six shillings in New England; eight shillings in New York; and 7s. 6d. in New Jersey. The English guinea is equal to 28s. in New England currency. The dollar will be considered 6s. in this book, unless notice is given of a different value.

6. How many pence are there in 84 farthings?

7. How many lb. of sugar, at 9d. per lb., may be bought for 117d.

8. How much beef, at 8 cents per lb., may be bought for $1.12 ?

9. How many lb. of steel, at 13 cents per lb., may be bought for $2.21 ?

10. How many cwt. of sugar, at $14 per cwt., may be bought for $280 ?

11. How many cwt. of cocoa, at $17 per cwt., may be bought for $391 ?

12. How much cocoa, at $25 per cwt., may be bought for 475 dollars ?

13. How much sugar, at 8d. per lb., may be bought for 4s. 8d. ?

14. How much cloth, at 4s. per yard, may be bought for 1£. 12s. ?

15. How much snuff, at 2d. 2 qr. per oz., may be bought for 40 farthings ?

16. How much wheat, at 8s. per bushel, may be bought for 2£. 16s. ?

17. How much cloth, at 7s. per yard, may be bought for 3£. 17s.

18. How much pork, at 9d. per pound, may be bought for 1£. 4s. 9d. ?

19. How much molasses, at 11d. per quart, may be bought for 2£. 15s. 11d.

20. In 38 shillings how many pounds ?

21. In 53 shillings how many pounds ?

22. In 87 shillings how many pounds ?

23. In 115 shillings how many pounds ?

24. In 178 shillings how many pounds ?

25. In 253 shillings how many pounds ?

26. In 6,247 shillings how many pounds ?

27. In 38 pence how many shillings ?

28. In 153 pence how many shillings ?

29. In 1,486 pence how many shillings ?

30. In 26,842 pence how many shillings ?

31. In 89 farthings how many pence ?

32. In 243 farthings how many pence ?

33. In 3,764 farthings how many pence ?

34. In 137 farthings how many pence ? How many shillings ?

35. In 382 farthings how many shillings ?

36. In 370 pence how many shillings? How many pounds?

37. In 846 pence how many pounds?

38. In 3,858 pence how many pounds?

39. In 2,340 farthings how many pence? How many shillings? How many pounds?

40. In 87,253 farthings how many pounds?

41. In 87 pints how many quarts? How many gallons?

42. In 230 pints how many gallons?

43. In 98 gills how many pints? How many quarts?

44. In 183 gills how many pints? How many quarts? How many gallons?

45. In 4,217 gills how many quarts? How many gallons?

46. In 23,864 gills how many gallons?

47. In 148 gallons how many hogsheads?

48. In 3,873 gallons how many pipes? How many tuns?

49. In 48,784 gills of wine how many hogsheads? How many pipes? How many tuns?

50. In 873 seconds how many minutes?

51. In 87 hours how many days?

52. In 73 days how many weeks? How many months?

53. In 2,738 minutes how many hours? How many days?

54 In 24,796,800 seconds how many minutes? How many hours? How many days? How many weeks? How many months?

55. In 506,649,600 seconds how many years, allowing 365 days to the year?

56. In 273 drams how many pounds Avoirdupois?

57. In 5,079 drams how many ounces? How many pounds?

58. In 573,440 drams how many ounces? How many pounds? How many quarters? How many hundred-weight? How many tons?

59. In 5,592,870 ounces how many tons?

60. In 384 grains Troy how many penny-weights?

61. In 325 dwt. how many ounces?

62. In 431 oz. Troy how many pounds?

63. In 198,706 grains Troy how many penny-weights? How many ounces? How many pounds?

64. In 678,418 grains Troy how many pounds?

65. In 37 nails how many yards?

66. In 87 nails how many ells English?

67. In 243 nails how many yards?

68. In 372 quarters how many ells Flemish?

69. In 3,107 nails how many ells Flemish?

70. In 327 shillings how many English guineas?

71. In 68 pence how many six-pences?

72. In 130 pence how many eight-pences?

73. In 342 pence how many four-pences?

74. In 2,086 pence how many nine-pences?

75. In 3,876 half-pence how many pence?

76. In 3,948 farthings how many pence? How many three-pences?

77. In 58,099 half-pence how many pounds?

78. In 57,604 farthings how many guineas at 28s. each?

79. In 3£. how many pence? How many three-pences?

80. In 73£. how many shillings? In these shillings how many dollars?

81. In 84£. how many shillings? In these shillings how many guineas?

82. In 37£. 4s. how many shillings? How many dollars?

83. How many pence are there in a dollar?

84. In 382 pence how many dollars?

85. In 32£. 8s. 4d. how many dollars?

86. In 13 yards how many quarters? In these quarters how many ells Flemish?

87. In 2 y. 3 qr. how many quarters? In these quarters how many ells English?

88. In 17 ells Flemish how many quarters? In these quarters how many aunes?

89. In 73 aunes how many yards?

90. From Boston to Liverpool is about 3,000 miles; if a ship sail at the rate of 115 miles in a day, in how many days will she sail from Boston to Liverpool?

91. If an ingot of silver weigh 36 oz. 10 dwt. how many pence is it worth at 3d. per dwt.? How many pounds?

92. How many spoons, weighing 17 dwt. each, may be made of 3lb. 6 oz. 18 dwt. of silver?

93. A goldsmith sold a tankard for 10£. 8s. at the rate of 5s. 4d. per ounce. How much did it weigh?

94. How many coats may be made of 47 yds. 1 qr. of broadcloth, allowing 1 yd. 3 qrs. to a coat?

95. What number of bottles, containing 1 pt. 2 gls. each, may be filled with a barrel of cider?

96. How many vessels, containing pints, quarts, and two

quarts, and of each an equal number, may be filled with a pipe of wine ?

Note. Three vessels, the first containing a pint, the second a quart, and the third two quarts, are the same as one vessel containing 3 qts. 1 pt. The question is the same as if it had been asked, how many vessels, each containing 3 qts. 1 pt., might be filled.

97. A man hired some labourers, men and boys, and of each an equal number ; to the men he gave 7s. and to the boys 3s. a day, each. How many shillings did it take to pay a man and a boy ? It took 3£. 10s. to pay them for 1 day's work. How many were there of each sort ?

Note. The question is the same as if it were asked, how many men this money would pay at 10s. per day.

98. A man bought some sheep and some calves, and of each an equal number, for $165.00 ; for the sheep he gave $7.75 apiece, and for the calves $3.25. How many were there of each sort ?

99. A man having $70.15, wished to purchase some rye, some wheat, and some corn, and an equal number of bushels of each kind. The rye was $0.95 per bushel, the wheat $1.37, and the corn $0.73. How many bushels of each sort could he buy if he laid out all his money ?

100. How many table spoons, weighing 23 dwt. each, and tea spoons, weighing 4 dwt. 6 gr. each, and of each an equal number, may be made from 4lb. 1 oz. 1 dwt. of silver ?

101. A merchant has 20 hhds. of tobacco, each containing 8 cwt. 3 qrs. 14 lb. which he wishes to put into boxes containing 7lb. each. How many boxes must he get ?

102. Bought 140 hhds of salt, at $4.70 per hhd. ; how much did it come to ? How many quintals of fish, at $2.00 per quintal, will it take to pay for it ?

103. A man bought 18 cords of wood, at 8 dollars a cord, and paid for it with flour, at $6 a barrel. How many barrels did it take ?

104. A man sold a hogshead of molasses at $0.40 per gal., and received his pay in corn at $0.84 per bushel. How many bushels did he receive ?

105. How much coffee, at $0.25 a pound, can I have for *100 lb. of tea,* at $0.87 per lb. ?

106. How much broadcloth, at $6.66 per yard, must be given for 2 hhds. of molasses, at $0.37 per gal. ?

107. How many times is 8 contained in 6,848 ?

108. 12,873 is how many times 3 ?

109. 86,436 is how many times 9 ?

110. 1,740 is how many times 6 ?

111. 18,345 is how many times 5 ?

112. 64,848 is how many times 4 ?

113. 94,456 is how many times 8 ?

114. 80,055 is how many times 15 ?

115. 8,772 is how many times 12 ?

116. 1,924 is how many times 37 ?

117. 1,924 is how many times 52 ?

118. 3,102 is how many times 94 ?

119. 3,102 is how many times 33 ?

120. 4,978 is how many times 131 ?

121. 28,125 is how many times 375 ?

122. 15,341 is how many times 529 ?

123. 49,640 is how many times 136 ?

124. 6,816,978 is how many times 8,253 ?

125. 92,883,780 is how many times 9,876 ?

126. 2,001,049,068 is how many times 261,986 ?

127. 11,714,545,304 is how many times 87,362 ?

128. 921,253,442,978,025 is how many times 918,273,645 ?

Miscellaneous Examples.

1. At 4s. 3d. per bushel, what cost 3 bushels of corn ?

2. At 2s. 3d. per yard, what cost 4 yards of cloth ?

3. What cost 7 lb. of coffee, at 1s. 6d. per lb. ?

4. What cost 3 gallons of wine, at 8s. 3d. per gal. ?

5. What cost 4 quintals of fish, at 13s. 3d. per quintal ?

6. What cost 5 cwt. of iron, at 1£. 9s. 4d. per cwt. ?

7. What cost 6 cwt. of sugar, at 3£. 8s. 4d. per cwt. ?

8. hat cost 9 yds. of broadcloth, at 2£. 6s. 8d. per yard ?W

9. How much sugar in 3 boxes, each box containing 14 lb. 7 oz. ?

10. At 3£. 9s. per cwt. what cost 7 cwt. of wool ?

11. What is the value of 5 cwt. of raisins, at 2£. 1s. 8d. per cwt. ?

12. How much wool in 3 packs, each pack weighing 2 cwt. 2 qrs. 13 lb. ?

4

13. What is the weight of 5 casks of raisins, each cask weighing 2 cwt. 3 qrs. 25 lb. ?

14. What is the weight of 12 pockets of hops, each pocket weighing 1 cwt. 2 qrs. 17 lb. ?

15. What is the weight of 16 pigs of lead, each pig weighing 3 cwt. 2 qrs. 17 lb. ?

Note. Divide the multiplier into factors as in Art. IV. ; that is, find the weight of 4 pigs and then of 16.

16. At 7s. 4d. per bushel, what cost 18 bushels of wheat ?

17. What cost 21 cwt. of iron, at 1£. 6s. 8d. per cwt. ?

18. What cost 28 lb. of tea, at 5s. 7d. per lb. ?

19. What cost 32 lb. of coffee, at 1s. 8d. per lb. ?

20. What cost 23 lb. of tea, at 4s. 3d. per lb. ?

Note. Find the price of 21 lb. and then of 2 lb. and add them together, Art. IV.

21. What cost 26 yds. of cloth, at 8s. 9d. per yd. ?

22. What cost 34 cwt. of rice, at 1£. 1s. 8d. per cwt. ?

23. If an ounce of silver cost 6s. 9d., what is that per lb. Troy ? What would 2 lb. 7 oz. cost ?

24. What is the value of 38 yds. of cloth, at 2£. 6s. 4d. per yd. ?

25. A man bought a bushel of corn for 5s. 3d., and a bushel of wheat for 7s. 6d. ; what did the whole amount to ?

26. How much silver in 6 table spoons, each weighing 5 oz. 10 dwts. ?

27. A man bought two loads of hay, one weighing 18 cwt. 3 qrs., and the other 19 cwt. 1 qr. ; how much in both ?

28. A man bought one load of hay for 7£. 3s., and another for 6£. 8s. 4d. ; how much did he give for both ?

29. A man bought 3 vessels of wine ; the first contained 18 gallons ; the second 15 gals. 3 qts. ; and the third 17 gals. 2 qts. 1 pt. How much in the 3 vessels ?

30. A merchant bought 4 pieces of cloth. The first contained 18 yds. 3 qrs. ; the second 23 yds. 1 qr. 3 nls. ; the third 25 yds. ; and the fourth 16 yds. 2 qrs. 2 nls. How many yards in the whole ?

31. A man bought 3 bu. 2 pks. of wheat at one time ; 18 bu. 3 pks. at another time ; 9 bu. 1 pk. 5 qts. at a third ; and 16 bu. 0 pk. 7 qts. at a fourth. How many bushels did he buy in the whole ?

32. A man bought a cask of raisins for 1£. 18s. 4d. ; 1 lb. of coffee for 1s. 6d. ; 1 cwt. of cocoa for 3£. 17s. ; 1 keg

of molasses for 13s. 7d. ; 1 box of lemons for 1£. 3s. ; 1 bushel of corn for 4s. 3d. How much did the whole amount to ?

33. A man bought 4 bales of cotton. The first contained 4 cwt. 2 qrs. 16 lb. ; the second 3 cwt. 1 qr. 14 lb. ; the third 5 cwt. 0 qr. 23 lb. ; and the fourth 4 cwt. 3 qrs. What was the weight of the whole ?

34. A merchant bought a piece of cloth, containing 19 yds. 3 qrs., and sold 4 yds. 1 qr. of it ; how much had he left ?

35. A grocer drew out of a hhd. of wine 17 gals. 3 qts. ; how much remained in the hogshead ?

36. A bought of B a bushel of wheat for 7s. 6d. He gave him 1 bushel of corn worth 5s. 3d. and paid the rest in money. How much money did he pay ?

37. C bought of B a bale of cotton for 18£. 4s. and B bought of C 4 barrels of flour for 9£. 3s. C paid B the rest in money. How much money did he pay ?

38. If from a piece of cloth, containing 9 yds. you cut off 1 yd. 1 qr., how much will there be left ?

39. If from a piece of cloth, containing 18 yds. 1 qr. you cut off 3 yds. 3 qrs., how much will be left ?

40. If from a box of butter, containing 15 lb. there be taken 6lb. 3 oz., how much will be left ?

41. A man sold a box of butter for 17s. 4d., and in pay received 7 lb. of sugar, worth 9d. 2qr. per lb. and the rest in money. How much money did he receive ?

42. A countryman sold a load of wood for 2£. 8s. and received in pay 3 gals. of molasses at 2s. 3d. per gal., 8 lb. of raisins at 10d. per lb., 1 gal. of wine at 11s. 3d., and the rest in money. How much money did he receive ?

43. A smith bought 17 cwt. 3 qrs. of iron, and after having wrought a few days, wishing to know how much of it he had wrought, he weighed what he had left, and found he had 8 cwt. 1 qr. 13 lb. How much had he wrought ?

44. A merchant bought 110 bars of iron, weighing 53 cwt. 1 qr. 11 lb., of which he sold 19 bars, weighing 9 cwt. 3 qrs. 15 lb. How much had he left ?

45. A merchant bought 17 cwt. 2 qrs. 1 lb. of sugar, and sold 13 cwt. 3 qrs. 17 lb. How much remains unsold ?

46. From a piece of cloth, which contained 43 yds. 1 qr., a tailor cut 3 suits, containing 6 yds. 2 qrs. 2 nls. each. How much cloth was there left ?

47. The revolutionary war between England and America commenced April 19th, 1775, and a general peace took place Jan. 20th, 1783. How long did the war continue ?

48. The war between England and the United States commenced June 18th, 1812, and continued 2 years 8. months 18 days. When was peace concluded ?

49. The transit of Venus (that is, Venus appeared to pass over the sun) A. D. 1769, took place at Greenwich, Eng. June 4th, 5 h. 20 min. 50 sec. morn. Owing to the difference of longitude between London and Boston it would take place 4 hours 44 min. 16 sec. earlier by Boston time. At what time did it take place at Boston ?

X.   1.* If 1 yard of cloth is worth 2 dollars, what is $\frac{1}{2}$ of a yard worth ?

2. What is $\frac{1}{2}$ of 2 dollars ?

3. If 2 dollars will buy 1 lb. of indigo, how much will 1 dollar buy ? How much will 3 dollars buy ? How much will 7 dollars buy ? How much will 23 dollars buy ? How much will 125 dollars buy.

4. At 3 shillings per bushel, what will $\frac{1}{3}$ of a bushel of corn cost ? What will $\frac{2}{3}$ of a bushel cost ?

5. At 3 dollars a barrel, what part of a barrel of cider will 1 dollar buy ? What part of a barrel will 2 dollars buy ? How much will 4 dollars buy ? How much will 5 dollars buy ? How much will 8 dollars buy ? How much will 28 dollars buy ?

6. At 3 dollars a box, how many boxes of raisins may be bought for 125 dollars ?

7. How many bottles, holding 3 pints each, may be filled with 85 gallons of cider ?

8. At 4 dollars a yard, how much will $\frac{1}{4}$ of a yard of cloth cost ? How much will $\frac{2}{4}$ of a yard cost ? How much will $\frac{3}{4}$ of a yard cost ?

9. A 4 dollars a box, what part of a box of oranges may be bought for 1 dollar ? What part for 2 dollars ? What part for 3 dollars ? How many boxes may be bought for 5 dollars ? How many for 19 dollars ?

10. At 4 dollars a barrel, how many barrels of rye flour may be bought for 327 dollars ?

11. At 5 dollars a cord, what will $\frac{1}{5}$ of a cord of wood

* See First Lessons, sect. III. art. B.

cost? What will $\frac{2}{5}$ cost? What will $\frac{3}{5}$ cost? What will $\frac{4}{5}$ cost? What will $\frac{6}{5}$ cost? What will $\frac{8}{5}$ cost?

12. At 5 dollars a week, what part of a week's board can I have for 1 dollar? What part for 2 dollars? What part for 3 dollars? What part for 4 dollars? How long can I be boarded for 7 dollars? How long for 18 dollars? How long for 39 dollars?

13. At 5 dollars a barrel, how many barrels of fish may be bought for $453?

14. If a firkin of butter cost 6 dollars, how much will $\frac{1}{6}$ of a firkin cost? How much will $\frac{2}{6}$ cost? How much will $\frac{5}{6}$ cost? How much will $\frac{7}{6}$ cost? How much will $\frac{11}{6}$ cost?

15. At 6 dollars a ream, what part of a ream of paper may be bought for 1 dollar? What part for 2 dollars? What part for 5 dollars? How many reams may be bought for 17 dollars? How many will 56 dollars buy?

16. At 6 dollars a barrel, how many barrels of flour may be bought for 437 dollars?

17. If a stage runs at the rate of 7 miles in an hour, in what part of an hour will it run 1 mile? In what part of an hour will it run 3 miles? In what part of an hour will it run 5 miles? In what time will it run 17 miles? In what time will it run 59 miles? In what time will it run from Boston to New York, it being 250 miles?

18. At 8 dollars a chaldron, how many chaldrons of coals may be bought for 75 dollars?

19. At 5 dollars a ream, how many reams of paper may be bought for 253 dollars?

20. At 7 dollars a barrel, how many barrels of flour may be bought for 2,434 dollars?

21. At 9 dollars a barrel, how many barrels of beef may be bought for 3,827 dollars?

22. At 8 dollars a cord, how many cords of wood may be bought for 853 dollars?

23. At 17 cents per lb., how many pounds of chocolate may be bought for $1.00? How many lb. for $2.00? How many lb. for $8.87?

24. At 25 dollars per cwt. what part of 1 cwt. of cocoa may be bought for 1 dollar? What part for 3 dollars? What part for 8 dollars? What part for 18 dollars? How many cwt. may be bought for 2,387 dollars?

25. At 28 dollars per ton, how many tons of hay may be bought for $427?

26. If 32 dollars will buy 1 thousand of staves, what part of a thousand may be bought for 1 dollar? What part of a thousand may be bought for 2 dollars? What part of a thousand may be bought for 7 dollars? What part for 15 dollars? What part for 27 dollars? How many thousands may be bought for 87 dollars? How many for $853?

27. At 45 cents per gallon, what part of a gallon may be bought for 1 cent? What part for 3 cents? What part for 8 cents? What part for 17 cents? What part for 37 cents? What part for 42 cents? How many gallons may be bought for $17.53?

28. At 138 dollars per ton, what part of a ton of potash may be bought for 1 dollar? What part for 17 dollars? What part for 35 dollars? What part for 87 dollars? What part for 115 dollars? How many tons may be bought for $875? How many tons for $27,484?

29. At $6.75 per barrel, what part of a oarrel of flour may be bought for 1 cent? What part for 17 cents? What part for 87 cents? What part for $2.87? How many barrels may be bought for $73.25?

30. At 73 cents a gallon, how many gallons of wine may be bought for $35.00?

31. At $2.75 per cwt., how many cwt. of fish may be bought for $93.67?

32. If a ship sail at the rate of 132 miles in a day, in how many days will she sail 3,000 miles?

33. If a ship sail at the rate of 125 miles per day, how long will it take her to sail round the world, it being about 24,911 miles?

34. How much indigo, at 2 dollars per lb., must be given for 19 yds. of broadcloth, at 7 dollars per yard?

35. How many bushels of corn, at 5s. per bushel, must be given for 23 bushels of wheat, at 7s. per bushel?

36. How many lb. of butter, at 23 cents per lb. must be given for 5 quintals of fish, worth $2.25 per quintal?

37. How many bushels of potatoes, at 3s. per bushel, must be given for a barrel of flour, worth 7 dollars and 4 shillings?

38. At 2£. 3s. per barrel, how many shillings will 7 barrels of flour come to? How much brandy, at 8s. per gal., will it take to pay for it?

39. If 63 gallons of water, in 1 hour, run into a cistern containing 423 gallons, in what time will it be filled?

40. At 4s. 3d. per bushel, what part of a bushel will 1d. buy ? What part of a bushel will 8d. buy ? What part of a bushel will 1s. or 12d. buy ? How many bushels may be bought for 2£. 16s. 4d ?

41. At 8s. 4d. per gallon, how many gallons of wine may be bought for 17£. 3s. 8d. ?

42. At 11s. 6d. per gallon, how many gallons of brandy may be bought for 43£. ?

43. A buys of B 3 cwt. 3 qrs. of sugar, at 9 cents per lb. ; 2 hhds. of brandy at $1.57 per gallon ; and 8 qqls. of fish at $2.55 per qql. In return, B pays A $25.00 in cash ; 150 lb. of bees-wax, at $0.40 per lb. ; and the rest in flour at $7.50 per barrel. How many barrels of flour must B give A ?

44. 785 are how many times 4 ?

45. 2,873 are how many times 8 ?

46. 8,467 are how many times 9 ?

47. 2,864 are how many times 14 ?

48. 43,657 are how many times 28 ?

49. 27,647 are how many times 78 ?

50. 884,673 are how many times 153 ?

51. 181,700 are how many times 437 ?

52. 984,607 are how many times 2,467 ?

53. Divide 1,708,540 by 13,841.

54. Divide 407,648,205 by 403,006.

55. Divide 100,000,000 by 12,478.

XI. 1. At 10 cents per lb., how many lb. of beef may be bought for $0.87 ?

2. At 10 cents per lb. how many lb. of cheese may be bought for $3.54 ?

3. At 10d. per lb. how many lb. of raisins may be bought for 13s. 4d. ?

4. Suppose you had 243 lb. of candles, which you wished to put into boxes containing 10 lb. each ; how many boxes would they fill ?

5. At 10 dollars a chaldron, how many chaldrons of coal may be bought for 749 dollars ?

6. At $1.00 per bushel, how many bushels of corn can you buy for $43.73 ?

7. If you had 32,487 oranges, which you wished to put into boxes containing 100 each, how many boxes could you fill ?

8. At $1.00 per lb. how many lb. of hyson tea may be bought for $243.84 ?

9. At $10.00 per bbl. how many barrels of pork may be bought for $247.63 ?

10. At $100.00 per ton, how many tons of iron may be bought for $8,734.87 ?

11. In 78 how many times 10 ?

12. In 3,876 how many times 10 ?

13. In 473 how many times 100 ?

14. In 6,783 how many times 100 ?

15. In 48,768 how many times 100 ?

16. In 475,384 cents how many dollars ?

17. In 5,710,648 how many times 1,000 ?

18. In 1,764,874 mills how many cents ? How many dimes ? How many dollars ?

19. In 4,710,074 mills how many dollars ?

XII.   1. What part of 5 lb. is 3 lb. ?

2. What part of 7 yards is 4 yards ?

3. What part of 7 yards is 10 yards ?

4. What part of 3 yards is 5 yards ?

5. What part of 4 oz. is 7 oz. ?

6. What part of 7d. is 10d. ?

7. What part of 17 cents is 9 cents ?

8. What part of 9 cents is 17 cents ?

9. What part of 35 dollars is 17 dollars ?

10. What part of 17 dollars is 35 dollars ?

11. 4 dollars is what part of 67 dollars ?

12. 67 dollars is what part of 4 dollars ?

13. What part of 103 rods is 17 rods ?

14. What part of 17 rods is 103 rods ?

15. What part of 256 miles is 39 miles ?

16. What part of 39 miles is 256 miles ?

17. What part of 287 inches is 138 inches ?

18. What part of 38,649 farthings is 8,473 farthings ?

19. What part of 907,384 is 3,906 ?

20. What part of 384 is 96,483 ?

21. What part of 1d. is 1 farthing ? What part of 1d. is 2 farthings ? 3 farthings ?

22. What part of 1s. is 1d. ? 2d. ? 3d. ? 4d. ? 5d. ? 6d. ? 7d. ? 11d. ?

23. What part of 1s. is 1 farthing ? 2 farthings ? 3 farthings ? 7 farthings ? 13 farthings ? 35 farthings ?

24. What part of 1s. is 1d. 3 qr. ? 2d. 1qr. ? 9d. 2qr. ?
Note. Reduce the pence to farthings.

25. What part of 1£. is 1 shilling ? 2 shillings ? 7 shillings ? 17 shillings ?

26. What part of 1£. is 1 penny ? 3 pence ? 7 pence ? 25 pence ? 87 pence ? 147 pence ?

27. What part of 1£. is 2s. 5d. ?

Note. Reduce the shillings to pence.

28. What part of 1£. is 7s. 4d ?

29. What part of 1£. is 13s. 8d. ?

30. What part of 1£. is 18s. 11d. ?

31. How many farthings are there in 1£. ?

32. What part of 1£. is 1 farthing ? 3 farthings ? 7 farthings ? 18 farthings ? 53 farthings ? 137 farthings ? 487 farthings ?

33. What part of 1£. is 7d. 3qr. ?

34. What part of 1£. is 11d. 2 qr. ?

35. What part of 1£. is 4s. 7d. 1 qr. ?

Note. Reduce the shillings and pence to farthings.

36. What part of 1£. is 13s. 8d. 2qr. ?

37. What part of a gallon is 1 quart ?

38. What part of a gallon is 1 pint ?

39. What part of a gallon is 1 gill ?

40. What part of a gallon is 7 gills ?

41. What part of a gallon is 2 qts. 1 pt. 3 gls. ?

42. What part of 1 hhd. is 1 gallon ? 17 gallons ?

43. What part of 1 hhd. is 1 gill ? 43 gills ?

44. What part of 1 hhd. is 17 gals. 3 qts. 1 pt. 2 gills ?

45. What part of 1 qr. is 1 lb. ? 13 lb. ?

46. What part of 1 lb. is 1 oz. Avoirdupois? 11 oz. ?

47. What part of 1 lb. is 1 dram ? 15 drams ?

48. What part of 1 lb. is 13 oz. 11 dr. ?

49. What part of 1 qr. is 1 dram ? 43 drams ?

50. What part of 1 qr. is 17 lb. 11 oz. 8 dr. ?

51. What part of 1 year is 1 calendar month ? 7 months ? 11 months ?

52. What part of a calendar month is 1 day ? 3 days ? 17 days ?

53. What part of 1 hour is 1 minute ? 17 minutes ?

54. What part of 1 day is 1 minute ? 13 minutes ?

55. What part of 1 day is 7 h. 43 min. ?

56. What part of 1 day is 1 second ? 73 seconds ? 258 seconds ?

57. What part of 1 day is 13 h. 43 min. 57 sec. ?

58. What part of a year is 1 second, allowing 365 days 6 hours to the year ? 8,724 seconds ?

59. What part of a year is 123 d. 17 h. 43 min. 25 sec. ?

60. What part of 8s. 3d. is 1 penny ? 8 pence ? 3s. 4d. ?

61. What part of 16s. 9d. is 5s. 3d. ?

62. What part of a dollar is 43 cents ?

63. What part of 5 dollars is 72 cents ?

64. What part of 3£. is 1 shilling ? 17 shillings ?

65. What part of 5£. is one penny ? 11 pence ? 4s. 8d. ?

66. What part of 4£. 7s. 8d. is 13s. 6d. ?

67. What part of 13£. 8s. 5d. is 3£. 7s. 6d. ?

68. What part of 3 yards is 1 quarter of a yard ?

69. What part of 16 yds. 1 qr. is 7 yds. 3 qrs. ?

70. What part of 13 yds. 3 qrs. 1 nl. is 4 yds. 3 qrs. 3 nls. ?

71. What part of 2 yds. 3 qrs. is 7 yds. 2 qrs. ?

72. What part of 3 days is 5 minutes ?

73. What part of 18 d. 3 h. is 13 d. 4 h. ?

74. What part of 5 d. 13 h. 18 min. is 26 d. 4 h. 7 min. ?

75. What part of 43 gals. 3 qts. 1 pt. is 27 gals. 2 qts. ?

76. What part of 17 gals. 1 qt. is 87 gals. 2 qts. ?

77. What part of 2 cwt. 1 qr. 17 lb. is 1 cwt. 3 qrs. 19 lb. ?

78. What is the ratio of 8 to 5 ?

79. What is the ratio of 5 to 8 ?

80. What is the ratio of 28 to 9 ?

81. What is the ratio of 9 to 28 ?

82. What is the ratio of 117 to 96 ?

83. What is the ratio of 57 to 294 ?

84. What is the ratio of 3,878 to 943 ?

XIII.   1.* If a family consume ⅓ of a barrel of flour in a week, how many barrels will last them 4 weeks ? How many barrels will last them 17 weeks ?

2. If ¼ of a barrel of cider will serve a family 1 week, how many barrels will serve them 11 weeks ? How many barrels will serve them 28 weeks ?

3. In ⅘ how many times 1 ? In ⅞ how many times 1 ?

---

* See First Lessons, Sect. VIII. Art. B.

4. If $\frac{1}{13}$ of a chaldron of coals will supply a fire 1 day, how many chaldrons will supply it 57 days at that rate?

5. Reduce $\frac{47}{13}$ to a mixed number.

6. In $\frac{64}{17}$ of a bushel how many bushels?

7. Reduce $\frac{64}{17}$ to a mixed number.

8. In $\frac{387}{20}$ of $\pounds$ how many pounds?

Note. This question is the same as the following.

9. In 387 shillings how many pounds?

10. In $\frac{437}{12}$ of a shilling how many shillings?

11. In 437 pence how many shillings?

12. In $\frac{134}{16}$ of a pound Avoirdupois, how many pounds?

13. In 134 oz. Avoirdupois how many pounds?

14. In $\frac{322}{28}$ of a guinea how many guineas?

15. In 322 shillings how many guineas, at 28 shillings each?

16. In $\frac{476}{24}$ of a day how many days?

17. In 476 hours how many days?

18. In $\frac{9737}{60}$ of an hour how many hours?

19. In 9,737 minutes how many hours?

20. In $\frac{43842}{365}$ of a year how many years?

21. In 43,842 days how many years, allowing 365 days to the year?

22. In $\frac{978468}{3847}$ of a year how many years?

23. Reduce $\frac{487}{45}$ to a mixed number.

24. Reduce $\frac{87563}{87}$ to a mixed number.

25. Reduce $\frac{3847}{784}$ to a mixed number.

26. Reduce $\frac{18006800}{24331}$ to a mixed number.

XIV. 1.* If $\frac{1}{7}$ of a cord of wood will supply two fires 1 day, how many days will a cord supply them? How many days will 3 cords supply them? How many days will 13 cords supply them?

2. How many 7ths are there in 1? How many 7ths are there in 3? How many in 13?

3. If $\frac{1}{8}$ of a barrel of beer will serve a family 1 day, how many days will 1 barrel serve them? How many days will $7\frac{1}{4}$ barrels serve them? How many days will $13\frac{3}{8}$ barrels serve them? How many days will $43\frac{5}{8}$ barrels serve them?

4. In 1 how many 8ths? In $7\frac{1}{4}$ how many 8ths? In $13\frac{3}{8}$ how many 8ths? In $43\frac{5}{8}$ how many 8ths?

5. If $\frac{1}{13}$ of a barrel of flour will serve a family 1 week,

---

* See First Lessons, Sect. VIII. Art. A.

how many weeks will $2\frac{4}{15}$ barrels serve them? How many weeks will $13\frac{7}{15}$ serve them?

6. In $13\frac{7}{15}$ how many 15ths?

7. If $\frac{1}{57}$ of a barrel of flour will serve 1 man 1 day, how many men will $7\frac{3}{57}$ barrels serve? How many men will $43\frac{35}{57}$ barrels serve?

8. Reduce $7\frac{3}{57}$ to an improper fraction.

9. Reduce $43\frac{35}{57}$ to an improper fraction.

10. In $13\frac{5}{8}$ bushels how many $\frac{1}{8}$ of a bushel?

11. In $23\frac{4}{17}$ barrels how many barrels?

12. In $4\frac{5}{12}$ shillings how many $\frac{1}{12}$ of a shilling? That is, in 4s. 5d. how many pence?

13. In $8\frac{7}{20}\pounds.$ how many $\frac{1}{20}$ of a pound? That is, in $8\pounds$ 7s. how many shillings?

14. In $15\frac{11}{24}$ days how many $\frac{1}{24}$ of a day?

15. In 15 d. 11 h. how many hours?

16. In $17\frac{43}{60}$ hours how many $\frac{1}{60}$ of an hour?

17. In 17 h. 43 min. how many minutes?

18. In $7\frac{37}{112}$ cwt. how many $\frac{1}{112}$ of 1 cwt.?

19. In 7 cwt. 37 lb. how many pounds?

20. In $18\frac{43}{237}$ cwt. how many $\frac{1}{237}$ of 1 cwt.?

21. In $237\frac{3}{15}$ how many $\frac{1}{15}$?

22. Reduce $437\frac{2}{11}$ to an improper fraction.

23. Reduce $63\frac{423}{2847}$ to an improper fraction.

XV. 1.* Bought 7 yards of cotton cloth, at $\frac{3}{5}$ of a dollar per yard; how many dollars did it come to?

2. If a horse consume $\frac{4}{7}$ of a bushel of oats in 1 day, how many bushels will he consume in 15 days?

3. If a family consume $\frac{2}{5}$ of a barrel of flour in a week, how many barrels would they consume in 17 weeks?

4. If $\frac{4}{7}$ of a ton of hay will keep 1 cow through the winter, how many tons will keep 23 cows the same time?

5. If a pound of beeswax cost $\frac{7}{20}$ of a dollar, how many dollars will 7 lb. cost?

6. If 1 lb of chocolate cost $\frac{4}{17}$ of a dollar, what will 27 lb. cost?

7. If one lb. of candles cost $\frac{3}{20}$ of a dollar, what will 43 lb. cost?

8. At $\frac{7}{25}$ of a dollar a pound, what cost 87 lb. of sheathing copper?

* See First Lessons, Sect. IX. Art. A.

9. At $\frac{17}{50}$ of a dollar a gallon, what will 1 hhd. of molasses cost ?

10. At $\frac{31}{100}$ of a dollar a gallon, what cost 3 hhds of molasses ?

11. At $\frac{89}{100}$ of a dollar a gallon, what cost 5 hhds of rum ?

12.* At $7\frac{1}{4}$ dollars per cwt. what cost 5 cwt. of lead ?

13. At $13\frac{1}{2}$ dollars per thousand, what cost 8 thousand of staves ?

14. At $14\frac{3}{8}$ dollars per barrel, what cost 23 barrels of fish ?

15. If a yard of cloth cost $38\frac{3}{37}$ shillings, what cost 15 yards ?

16. If a barrel of beef cost $54\frac{13}{87}$ shillings, what cost 23 barrels ?

17. If 1 gallon of gin cost $\frac{67}{253}$ of $1\pounds$. what cost 1 hhd. ?

18. At $2\frac{33}{987}\pounds$. per barrel, what cost 17 barrels of flour ?

19. A man failing in trade is able to pay only $\frac{3}{5}$ of a dollar on a dollar, how much will he pay on a debt of 5 dollars ? How much on 53 dollars ?

20. A man failing in trade is able to pay only $1\frac{3}{7}$ of a dollar on a dollar, how much will he pay on a debt of 75 dollars ? How much on a debt of 153 dollars ?

21. Suppose the duties on tea to be $\frac{28}{100}$ of a dollar on 1 lb., what would be the duties on 738 lb. ?

22. A man failing in trade is able to pay only $\frac{387}{155}$ of a dollar on a dollar, how much can he pay on a debt of 873 dollars ?

23. How much is 5 times $\frac{8}{17}$?

24. How much is 7 times $\frac{58}{215}$?

25. How much is 17 times $\frac{38}{275}$?

26. How much is 9 times $\frac{475}{2868}$?

27. How much is 35 times $\frac{215}{17835}$?

28. How much is 237 times $\frac{1168}{18638}$?

29. Multiply $\frac{17}{3876}$ by 238.

30. Multiply $\frac{28}{18405}$ by 1003.

31. Multiply $\frac{200}{17063}$ by 5060.

32. Multiply $\frac{16847}{1893700}$ by 607.

XVI. 1.† If a piece of linen cost 24 dollars, what will $\frac{1}{4}$ of a piece cost ?

2. If 3 chaldrons of coal cost 36 dollars, what part of 36

---

* See First Lessons, Sect. IX. Art. B.
† See First Lessons, Sect. V. and X.

dollars will 1 chaldron cost ? How much will a chaldron cost ?

3. If 7 lb. of chocolate cost $1.54, what part of $1.54 will 1 lb. cost ? What is $\frac{1}{7}$ of $1.54 ?

4. If 9 yards of cloth cost 126 dollars, what part of 126 dollars will 1 yard cost ? How much will it cost per yard ?

5. If 17 chaldrons of coal cost 136 dollars, what part of 136 dollars will 1 chaldron cost ? What is $\frac{1}{17}$ of 136 dollars ?

6. A ticket drew a prize of 652 dollars, of which A owned $\frac{1}{4}$; what was A's share of the money ?

7. A privateer took a prize worth 36,960 dollars, of which the captain was to have $\frac{1}{8}$, the first mate $\frac{1}{12}$, the second mate $\frac{1}{16}$, and the rest was to be divided equally among the crew, which consisted of 50 men; what was the share of each officer, and of each sailor ?

8. If a man travel 38 miles in a day, how far will he travel in $7\frac{1}{2}$ days ?

9. At $2.48 per barrel, what will $5\frac{1}{2}$ barrels of cider cost ?

10. At $1.38 a bushel, what will $8\frac{1}{2}$ bushels of rye cost ?

11. At $1.83 per bushel, what will $\frac{1}{3}$ of a bushel of wheat cost ? What will $\frac{2}{3}$ cost ?

12. At $7.23 per barrel, what cost $4\frac{1}{3}$ barrels of flour ?

13. At $1.92 per gallon, what cost $\frac{1}{4}$ of a gallon of brandy ? That is, what cost 1 quart ?

14. At $4.20 per box, what cost $\frac{1}{4}$ of a box of oranges ? What cost $\frac{3}{4}$ of a box ? What cost $1\frac{3}{4}$ box ?

15. At $2.20 per lb., what cost $\frac{3}{4}$ of a lb. of indigo ? What cost $7\frac{3}{4}$ lb. ?

16. At $2.25 per quintal, what cost $\frac{2}{3}$ of a qql. of fish ? What cost $11\frac{2}{3}$ qqls. ?

17. At $7.75 per cwt., what cost $\frac{1}{5}$ cwt. of sugar ? What cost $\frac{2}{5}$ cwt. ? What cost $\frac{3}{5}$ cwt. ?

18. At $7.25 per cask, what cost $3\frac{1}{5}$ casks of Malaga raisins ?

19. At $0.75 per bushel., what cost $18\frac{4}{5}$ bushels of Indian corn ?

20. At $6.78 per barrel, what cost $\frac{1}{6}$ of a barrel of flour ? What cost $\frac{5}{6}$ of a barrel ?

21. At $7.86 per barrel, what cost $18\frac{5}{6}$ barrels of flour ?

22. If 7 bushels of oats cost $2.94, what part of $2.94 will 1 bushel cost ? What is $\frac{1}{7}$ of $2.94 ?

23. A man bought 8 sheep for $60.24 ; what part of $60.24 did 1 sheep cost ? What is $\frac{1}{8}$ of $60.24 ?

24. A merchant bought 12 barrels of flour for $82.44; what part of $82.44 did 1 barrel cost? What is $\frac{1}{12}$ of $82.44?

25. A merchant bought 18 hhds. of brandy for $1,692.00; what part of $1,692.00 did 1 hhd. cost? What did it cost per hhd.?

26. If 37 lb. of beef cost $2.96, what part of $2.96 will 1 lb. cost? What is $\frac{1}{37}$ of $2.96?

27. If 1 hhd. of rum cost $52.92 what part of $52.92 will 1 gallon cost? How much will 1 gallon cost?

28. At 43 cents a gallon, what will 15$\frac{3}{4}$ hhds. of molasses come to?

29. How many inches are there in a mile?

MEASURE OF LENGTH.

3	barley-corns (bar.) make 1 inch, marked	in.
12	inches 1 foot	ft.
3	feet 1 yard	yd.
5$\frac{1}{2}$	yards or } 1 rod	rod.
16$\frac{1}{2}$	feet } or pole	pol.
40	poles 1 furlong	fur.
8	furlongs 1 mile	ml.
3	miles 1 league	l.
60	geographical miles, or } 1 degree nearly,	{ deg.
69$\frac{1}{2}$	statute miles }	or °
360	degrees the circumference of the earth.	

Also 4	inches make	1 hand
5	feet	1 geometrical pace
6	feet	1 fathom
6	points	1 line
12	lines	1 inch

30. How many geographical miles is it round the earth?
31. How many statute miles round the earth?
32. How many inches in 15 miles?
33. How many rods round the earth?
34. How many barley-corns will reach round the earth?
35. At $25.00 per ton, what will 1 cwt. of hay come to?
36. If 6 horses eat 18 bushels of oats in a week, what part of 18 bu. will 1 horse eat in the same time? What part of 18 bu. will 5 horses eat? What is $\frac{5}{6}$ of 18 bu.?
37. If a man travel 35 miles in 7 hours, how many miles

will he travel in 1 hour ? How many in 12 hours ? How many in 53 hours ?

38. If a stage run 96 miles in 12 hours, how many miles will it run in 15 days 5 hours, at that rate, if it run 12 hours each day ?

39. At $30.00 a ton, what will 7 tons 8 cwt. of hay come to ?

40. A man, after travelling 23 hours, found he had travelled 115 miles ; how far had he travelled in an hour, supposing he had travelled the same distance each hour ? how far would he travel in 47 hours at that rate ?

41. If 1 hhd. 20 gal. cost $118.69, what is it a gallon ? How much is it per hhd. ? How much would 3 hhds. 17 gal. come to, at that rate ?

42. If 18 gal. 3 qts. of wine cost $33.75, what is it a quart ? What will 1 hhd. 43 gals. 2 qts. come to, at that rate ?

43. If 3 qrs. 13 lb. of cocoa cost $14.55, what is it per lb. ? How much will 47 lb. come to, at that rate ?

44. If 1 cwt. 3 qr. 7 lb. of cocoa cost $32.48, what is it per lb. ? What would be the price of 3 cwt. 2 qrs. 5 lb. at that rate ?

45. If 1 oz. of silver be worth 6s. 8d., what is that per dwt. ? What would be the price of a silver cup, weighing 10 oz. 14 dwts. ?

46. If 1 cwt. 3 qrs. 23 lb. of tobacco cost $54.75, what will 3 cwt. 2 qrs. 5 lb. cost at that rate ?

47. If 6 horses will consume 19 bu. 2 pks. of oats in 3 weeks, how many pecks will 17 horses consume in the same time ? How many bushels ?

48. A ship was sold for £568, of which A owned ⅜ ; what was A's part of the money ?

49. If 3 yds. 3 qrs. of broadcloth cost $30.00, what will 7 yds. cost ?

50. If 37 yds. of cloth cost $185.00, what will 18¾ yds. cost ?

51. If 23 yds. of cloth cost $230.00, what will 1 qr. cost ? What will 1 ell English cost ? What will 17½ ells cost ?

52. If a chest of Hyson tea, weighing 79 lb., cost 32£. 11s. 9d., what would 43 lb. come to at that rate ?

53. If 9 cwt. 3 qrs. 4 lb. of tallow cost $109.60, what will 1 cwt. cost ?

54. If the distance from Boston to Providence be 40 miles,

how many times will a carriage wheel, the circumference of which is 15 ft. 6 in., turn round in going that distance?

55. If the forward wheels of a wagon are 14 ft. 6 in., and the hind wheels 15 ft. 9 in. in circumference, how many more times will the forward wheels turn round than the hind wheels, in going from Boston to New York, it being 248 miles?

56. How many times will a ship 97 ft. 6 in. long, sail her length in the distance of 1,200 miles?

57. If 1 bushel of oats will serve 3 horses 1 day, how much will serve 1 horse the same time? How much will serve 2 horses?

58. If 1 bushel of corn will serve 5 men 1 week, how much will serve 1 man the same time? How much will serve 3 men?

59. If you divide 1 gallon of beer equally among 5 men, how much would you give them apiece? If you divide 2 gallons, how much would you give them apiece? If you divide 3 gallons, how much would you give them apiece? If you divide 7 gallons, how much would you give them apiece?

60. What is $\frac{1}{5}$ of 1? What is $\frac{1}{5}$ of 2? What is $\frac{1}{5}$ of 3? What is $\frac{1}{5}$ of 7?

61. If 7 yards of cloth cost 1 dollar, what part of a dollar will 1 yard cost? If 7 yards cost 2 dollars, what part of a dollar would 1 yard cost? If 7 yards cost 5 dollars, what part of a dollar would 1 yard cost? If 7 yards cost 10 dollars, what part of a dollar will 1 yard cost? How many dollars?

62. What is $\frac{1}{7}$ of 1? What is $\frac{1}{7}$ of 2? of 3? of 5? of 10?

63. If you divide 1 gallon of wine equally among 13 persons, how much would you give them apiece? How much if you divide 2 gallons? How much if you divide 3 gallons? 5 gallons? 11 gallons? 15 gallons? 23 gallons? 57 gallons?

64. What is $\frac{1}{13}$ of 1? of 2? of 3? of 5? of 11? of 23? of 57?

65. If you divide 1 dollar equally among 23 persons, what part of a dollar would you give them apiece? If you divide 2 dollars, what part of a dollar would you give them apiece? 7 dollars? 18 dollars? 34 dollars? 87 dollars? 253 dollars?

66. What is $\frac{1}{23}$ of 1? of 2? of 7? of 18? of 34? of 87? of 253?

67. If 8 barrels of flour cost 53 dollars, what is that a barrel ?   What will 13 barrels cost ?

68. If 17 lb. of beef cost $1.43, what is that per lb. ?

69. If 57 lb. of raisins cost $8.37, how much is that per lb. ?   What will 43 lb. cost ?

70. If 1 cwt. 3 qrs. 15 lb. of sugar cost $19.53, how much is it per lb. ?   What will 6 cwt. 1 qr. 23 lb. cost ?

71. If 15 yds. 3 qrs. of broadcloth cost $147.23, what will 1 qr. cost ?   What will a yard cost ?   What will 57 yards cost ?

72. Bought 3 hhds. of wine for $257.00 ; what was it per gallon ?   What would 5 pipes cost at that rate ?

73. If 2 bushels of wheat is sufficient to sow 3 acres, what part of a bushel will sow 1 acre ?   How much will sow 5 acres ?

74. If 5 barrels of cider will serve 8 men 1 year, what part of a barrel will serve 1 man the same time ?   How much will serve 17 men ?

75. If 5 barrels of flour will serve 23 men 1 month, what part of a barrel will serve 1 man the same time ?   How much will serve 75 men ?

76. If 3 acres produce 43 bushels of wheat, what part of an acre will produce 1 bushel ?   How much will produce 7 bushels ?   How much will produce 28 bushels ?   How much will produce 153 bushels ?

77. If 7 acres 1 rood produce 123 bushels 3 pks. of wheat, how much will 1 rood produce ?   How much will 25 acres produce ? .

Note. 4 roods make 1 acre.

78. If 9 acres 1 rood produce 136 bushels of rye, what part of a rood will produce 1 bushel ?   How many acres will produce 500 bushels ?

79. If 435 men consume 96 barrels of provisions in 9 months, how many barrels will 2,426 men consume in the same time ?

80. At 23 cents per gallon, what will $\frac{3}{5}$ of a hhd. of molasses come to ?

81. At 14 cents per lb., what will $\frac{1}{4}$ of 1 cwt. of raisins come to ?

82. How many shillings in $\frac{3}{5}$ of 13s. ?

83. How many pence in $\frac{3}{4}$ of a shilling ?

84. How many pence in $\frac{4}{5}$ of a shilling ?

85. How many farthings in $\frac{3}{4}$ of a penny ?

86. Find the value of $\frac{1}{2}$ of a shilling, in pence and farthings.

87. Find the value of $\frac{4}{5}$ of a shilling, in pence and farthings.

88. Find the value of $\frac{2}{3}$ of 1£., in shillings and pence.

89. Find the value of $\frac{4}{5}$ of 1£., in shillings, pence, and farthings.

90. What is the value of $\frac{3}{13}$ of 1£., in shillings, pence, and farthings ?

 91. Find $\frac{4}{5}$ of a day in hours, minutes, and seconds.

 92. Find $\frac{3}{8}$ of 1 hour in minutes and seconds.

 93. What is $\frac{5}{13}$ of a day ?

 94. What is $\frac{7}{23}$ of a day ?

 95. What is $\frac{3}{8}$ of 1 lb. Avoirdupois ?

 96. What is $\frac{4}{7}$ of 1 cwt. in quarters and lb. ?

 97. What is $\frac{5}{17}$ of 1 cwt. ?

 98. What is $\frac{5}{18}$ of 1 hhd. of wine ?

 99. What is $\frac{4}{35}$ of 1 hhd. of wine ?

100. What is $\frac{5}{7}$ of a yard ?

101. What is $\frac{9}{11}$ of a yard ?

102. What is $\frac{5}{17}$ of a yard ?

103. What is $\frac{3}{7}$ of a dollar ?

104. What is $\frac{5}{28}$ of a dollar ?

105. What is $\frac{6}{47}$ of a dollar ?

106. What is $\frac{9}{23}$ of 1£. ?

107. What is $\frac{14}{34}$ of 1£. ?

108. What is $\frac{34}{73}$ of 1£. ?

109. What is $\frac{43}{47}$ of a gallon of wine ?

110. What is $\frac{27}{73}$ of a shilling ?

111. What is $\frac{38}{57}$ of a day ?

112. What is $\frac{37}{135}$ of a dollar ?

113. What is $\frac{43}{87}$ of a yard ?

114. What is $\frac{18}{43}$ of a bushel ?

115. What is $\frac{18}{37}$ of 1 lb. Avoirdupois ?

116. What is $\frac{23}{67}$ of 1£. ?

117. What is $\frac{83}{130}$ of 1£. ?

118. What is $\frac{123}{287}$ of 1£. ?

119. What is $\frac{248}{823}$ of 1 cwt. ?

120. What is $\frac{484}{1273}$ of a week ?

121. What is $\frac{248}{684}$ of 1 hhd. of brandy ?

122. What will $\frac{233}{477}$ of 1 hhd. of wine come to, at $1.23 per gal. ?

123. What will $\frac{187}{443}$ of 1 cwt. of sugar come to, at $0.12 per lb. ?

67. What will 4¾ tons of iron come to, at $4.00 per cwt. ?

125. What will 7 4/8 cwt. of sugar come to, at 8 cents per lb. ?

126. What will 8½ hhd. of molasses come to, at $0.48 per gal. ?

127. What will 19 1/3 tons of iron come to, at $0.05 per lb. ?

128. What will 23 1/7 pipes of brandy come to, at $1.43 per gal. ?

129. At 5s. per bushel, what will 4 bu. 3 pks. 5 qts. of corn come to ?

130. At $9.00 per cwt., what will be the price of 1lb. of sugar? What will 3 cwt. 2 qrs. 7 lb. come to at that rate ?

131. At $87.00 per cwt., what cost 4 chests of tea, each weighing 3 cwt. 3 qrs. 14 lb. ?

132. What cost 18 gals. 3 qts. of brandy, at the rate of $97.00 per hhd. ?

133. Bought a silver cup weighing 9oz. 4dwt. 16 grs. for 3£. 2s. 3d. How much was it per grain. How much per ounce ?

134. Bought a silver tankard weighing 1 lb. 8 oz. 17 dwt. 13 gr. for $25.00 ; how much was it per ounce ?

135. If 34 tons 9 cwt. 2 qrs. 18 lb. of tallow cost $6,500,00, what is it per lb. ? How much per ton ?

136. A and B traded ; A sold B 8 cwt. of sugar, at 12 cents per lb. ; how much did it come to ? In exchange, B gave A 18 cwt. of flour ; what was the flour rated at per lb. ?

137. B delivered C 2 pipes of brandy, at $1.40 per gallon, for which he received 87 yards of cloth ; what was the cloth valued at per yard ?

138. D sells E 370 yards of cotton cloth at 33 cents per yard ; for which he receives 500 lb. of pepper ; what does the pepper stand him in per lb. ?

139. A merchant bought 3 hhds. of brandy, at $1.30 per gallon, and sold it so as to gain ⅓ of the first cost ; how much did he sell it for per gallon ?

140. A merchant bought a quantity of tobacco for $250,00, and sold it so as to gain 3/10 of the first cost ; how much did he sell it for ?

141. A merchant bought 1 hhd of wine for $80.00 ; how much must he sell it for to gain $15.00 ? How much will that be a gallon ?

142. A merchant bought 500 barrels of flour for $2,000,00 ; how much must he sell it for per barrel to gain $250.00 on the whole ?

143. A merchant bought 350 yards of cloth for $1,800,00; how much must he sell it for to gain $\frac{3}{10}$ of the first cost? How much will that be a yard?

144. A merchant bought 2 hhds. of molasses for $35.28; how much must he sell it for per gal. to gain $\frac{3}{4}$ of the first cost?

145. A merchant bought 7 cwt. of coffee for $175.00, but being damaged he was willing to lose $\frac{1}{3}$ of the first cost. How much did he sell it for per lb.?

146. A merchant sold 7 cwt. of rice for $22.75, to receive the money in 6 months, but for ready money he agreed to make a discount of $\frac{3}{100}$ of the whole price. How much was the rice per lb. after the discount?

147. If 8 boarders will drink a cask of beer in 12 days, how long would it last 1 boarder? How long would it last 12 boarders?

148. If 23 men can build a wall in 32 days, how many men would it take to do it in 1 day? How many men will it take to do it in 8 days?

149. If 15 men can do a piece of work in 84 days, how many men must be employed to perform the whole in 1 day? How many to do it in 30 days?

150. If 18 men can perform a piece of work in 45 days, how many days would it take 1 man to do it? How long would it take 57 men to do it?

151. If 25 men can do a piece of work in 17 days, in how many days will 38 men do it?

152. If a man perform a journey in 8 days, by travelling 12 hours in a day, how many hours is he performing it? How many days would it take him to perform it if he travelled only 8 hours in a day?

153. If a man, by working 11 hours in a day, perform a piece of work in 24 days, how many days will it take him to do it if he works 13 hours in a day?

154. If I can have 5 cwt. carried 138 miles for 11 dollars, how far can I have 25 cwt. carried for the same money?

155. Suppose a man agrees to pay a debt with wheat, and that it will take 43 bushels to pay it, when wheat is 7 shillings per bushel; how much will it take when wheat is 9 shillings per bushel?

156. If 11 men can do a piece of work in 14 days, when the days are 15 hours long, how many men would it take to do it in the same number of days, when the days are 11 hours long?

157. If 5 men can do a piece of work in 5 months by working 7 hours in a day, in how many months will they do it, if they work 10 hours in a day?

158. Two men, A and B, traded in company; A furnished ⅖ of the stock and B ⅓; they gained $864.00; what was each one's share of the gain?

159. Three men, A, B, and C, traded in company; A furnished $\frac{17}{48}$ of the capital; B $\frac{19}{48}$, and C the rest. They gained $8,453,67; what was each one's share of the dividend?

160. Two men, B and C, bought a barrel of flour together. B paid 5 dollars and C 3 dollars; what part of the whole price did each pay? What part of the flour ought each to have?

161. Two men, C and D, bought a hogshead of wine; C paid $47.00, and D 53.00; how many dollars did they both pay? What part of the whole price did each pay? How many gallons of the wine ought each to have?

162. Three men, C, D, and E, traded in company; C put in $850.00; D, 942.00; and E, $1,187.00; how many dollars did they all put in? What part of the whole did each put in? They gained $1,353.18; what was each man's share of the gain?

163. Five men, A, B, C, D, and E, freighted a vessel: A put on board goods to the amount of $4,000.00; B, $15,000.00; C, $11,000.00; D, $7,500.00; and E, $850.00. During a storm the captain was obliged to throw overboard goods, to the amount of $13,400.00; what was each man's share of the loss?

164. Three men bought a lottery ticket for $20.00; of which F paid $4.37; G $8.53; and H, the rest. They drew a prize of $15,000.00; what was the share of each?

165. Three men hired a pasture for $42.00; the first put in 4 horses; the second, 6; and the third, 8. What ought each to pay?

166. A man failing in trade, owes to A $350.00; to B $783.00; to C $1,780.00; to D $2,875.00; and he has only $2,973.00 in property, which he agrees to divide among his creditors in proportion to the several debts. What will each receive?

167. What is $\frac{137}{2361}$ of 378,648?

168. What is $\frac{1344}{4444}$ of 87?

169. What is $\frac{143}{11111}$ of 3?

170. What is $\frac{111}{1133}$ of 47 ?

171. Multiply $\frac{211}{311}$ by 7.

172. What is $\frac{311}{611}$ of 7 ?

173. Multiply 973 by $\frac{311}{111}$.

174. Multiply $\frac{311}{111}$ by 973.

175. Multiply $47\frac{1}{4}$ by $\frac{15}{231}$.

176. Multiply $\frac{15}{231}$ by 471.

177. Multiply $\frac{967}{1000}$ by 138.

178. Multiply 138 by $\frac{967}{1000}$.

179. Multiply $\frac{24}{1303}$ by 950.

180. Multiply 950 by $\frac{24}{1303}$.

XVII.   1. If 2 lb. of figs cost $\frac{2}{3}$ of a dollar, what is that a pound ?

2. If 2 bushels of potatoes cost $\frac{4}{5}$ of a dollar, what is that a bushel ?   What would be the price of 8 bushels at that rate ?

3. If $\frac{2}{5}$ of a barrel of flour were to be divided equally among 3 men, how much would each have ?

4. If 3 horses consume $\frac{9}{17}$ of a ton of hay in 1 month, how much will 1 horse consume ?   How much would 11 horses consume in the same time ?

5.* If 3 lb. of beef cost $1\frac{2}{5}$ of a dollar, what would a quarter of beef, weighing 136 lb., cost at that rate ?

6. If 2 yds. of cloth cost $8\frac{3}{4}$ dollars, what will 7 yards cost at that rate ?

7. If 4 bushels of wheat cost $32\frac{4}{5}$ shillings, what will 17 bushels cost ?

8. If 3 sheep are worth $23\frac{4}{5}$ bushels of wheat, how many bushels is 1 sheep worth ?   How many bushels are 50 sheep worth at that rate ?

Note. Reduce $23\frac{4}{5}$ to fifths, or divide as far as you can, and then reduce the remainder to fifths, and take $\frac{1}{3}$ of them.

9. If 7 calves are worth $59\frac{1}{3}$ bushels of corn, how many bushels are 15 calves worth at that rate ?

10. A man laboured 15 days for $20\frac{4}{5}$ dollars ; how much would he earn in 3 months at that rate, allowing 26 working days to the month ?

11. A man travelled $88\frac{1}{11}$ miles in 17 hours ; how far did he travel in an hour ?

_____

* See First Lessons, Sect. XIV.

12. A man travelled $476\frac{1}{4}$ miles in 8 days; how far did he travel each day, supposing he travelled the same number of miles each day ?

13. Divide $77\frac{8}{11}$ bushels of corn equally among 15 men.

14. If 23 yards of cloth cost $175\frac{3}{4}$ dollars, what is that a yard ?

15. If 35 lb. of raisins cost $3\frac{85}{100}$ dollars, what will 2 cwt. cost at that rate ?

16. A man divided $\frac{1}{4}$ of a water-melon equally between 2 boys; how much did he give them apiece ?

17. Suppose you had $\frac{1}{4}$ of a pine apple and should divide it into two equal parts; what part of the whole apple would each part be ?

18. If you divide $\frac{3}{4}$ of a bushel of corn equally between 2 men, how much would you give them apiece ?

19. What is $\frac{1}{2}$ of $\frac{3}{4}$?

20. If you divide $\frac{1}{3}$ of a bushel of grain between two men, how much would you give them apiece ?

Note. Cut the third into two parts. What will the parts be ?

21. What is $\frac{1}{2}$ of $\frac{1}{3}$?

22. If you divide $\frac{1}{4}$ of a barrel of flour equally between two men, how much will you give them apiece ?

23. What is $\frac{1}{2}$ of $\frac{1}{4}$?

24. A man having $\frac{2}{3}$ of a barrel of flour divided it equally among 4 men; how much did he give them apiece ?

25. What is $\frac{1}{4}$ of $\frac{2}{3}$?

26. I₁ 3 lb. of sugar cost $\frac{2}{3}$ of a dollar, what is it a pound ?

27. What is $\frac{1}{3}$ of $\frac{2}{3}$?

28. If 5 lb. of rice cost $\frac{2}{3}$ of a dollar, what is that a pound ?

29. If 3 lb. of raisins cost $\frac{1}{2}$ of a dollar, what is that a pound ? What will 2 lb. cost at that rate ? What 7 lb. ?

30. What is $\frac{1}{2}$ of $\frac{1}{4}$? What is $\frac{2}{3}$ of $\frac{1}{4}$? What is $\frac{7}{3}$ of $\frac{1}{4}$?

31. If 7 lb. of sugar cost $\frac{2}{3}$ of a dollar, what is it a pound ? What will 5 lb. cost at that rate ? What would 15 lb. cost ?

32. What is $\frac{1}{4}$ of $\frac{2}{3}$? What is $\frac{4}{5}$ of $\frac{2}{3}$? What is $\frac{15}{7}$ of $\frac{2}{3}$?

33. During a storm, a master of a vessel was obliged to throw overboard $\frac{4}{15}$ of the whole cargo. What part of the whole loss must a man sustain who owned $\frac{3}{4}$ of the cargo ?

34. A man owned $\frac{3}{25}$ of the capital of a cotton manufactory, and sold $\frac{4}{11}$ of his share. What part of the whole capital did he sell ? What part did he then own ?

35. If 3 bushels of wheat cost $5\frac{1}{2}$ dollars, what is it a bushel? What will 2 bushels cost at that rate?

36. What is $\frac{1}{3}$ of $5\frac{1}{4}$? What is $\frac{2}{3}$ of $5\frac{1}{4}$?

37. If 4 dollars will buy $5\frac{3}{4}$ bushels of rye, how much will one dollar buy? How much will 3 dollars buy?

38. What is $\frac{1}{4}$ of $5\frac{2}{3}$? What is $\frac{3}{4}$ of $5\frac{2}{3}$?

39. If 17 barrels of flour cost $\$107\frac{2}{5}$, what will 23 barrels cost?

40. What is $\frac{23}{17}$ of $107\frac{2}{5}$?

41. If 12 cwt. of sugar cost $\$137\frac{3}{8}$, what is the price of 1 qr.? What of 1 lb.?

42. At 4 dollars for $3\frac{1}{7}$ gallons of wine, how much may be bought for $67\frac{1}{2}$ dollars?

Note. Find how much $\frac{1}{2}$ a dollar will buy.

43. If 3 cords of wood cost 20 dollars, what will $7\frac{1}{2}$ cords cost?

44. If 19 yards of cloth cost 155 dollars, what will be the price of $1\frac{7}{8}$ yards?

45. If 18 lb. of raisins cost $2\frac{2}{7}$ dollars, what is that per lb.? What would be the price of $5\frac{2}{3}$ lb. at that rate?

46. If 11 lb. of butter cost $2\frac{3}{20}$ dollars, what will $18\frac{2}{3}$ lb. cost?

47. If 7 gallons of vinegar cost $\frac{3}{4}$ of a dollar, what will $27\frac{1}{4}$ gallons cost?

48. If 1 lb. of sugar cost $\frac{11}{12}$ of a dollar, what will $17\frac{3}{4}$ lb. cost?

49. If a yard of cloth cost $7\frac{9}{10}$ dollars, what will $\frac{2}{3}$ of a yard cost?

50. At $\frac{4}{25}$ of a dollar a yard, what will $\frac{2}{3}$ of a yard of cloth cost?

51. At $3\frac{2}{3}$ shillings a yard, what will $7\frac{1}{3}$ yards of riband cost?

52. At 3 dollars a barrel, what part of a barrel of cider may be bought for $\frac{1}{2}$ of a dollar?

53. At 4 dollars a yard, what part of a yard of cloth may be bought for $\frac{1}{4}$ of a dollar?

54. At 2 dollars a yard, how much cloth may be bought for $5\frac{1}{2}$ dollars?

55. At 2 dollars a gallon, how much brandy may be bought for $7\frac{3}{4}$ dollars?

56. At 3 shillings a quart, how many quarts of wine may be bought for $17\frac{3}{4}$ shillings?

57. At 6 dollars a barrel, how many barrels of flour may be bought for 45$\frac{3}{11}$ dollars?

58. If 1 cwt. of iron cost 4$\frac{2}{3}$ dollars, what will 5$\frac{3}{8}$ cwt. cost?

59. A man failing in trade can pay only $\frac{2}{3}$ of a dollar on each dollar, how much can he pay on 7$\frac{1}{2}$ dollars? How much on 23$\frac{4}{5}$ dollars?

60. A man failing in trade is able to pay only $\frac{13}{15}$ of a pound on a pound, how much can he pay on 17£. 15s.?

61. A man failing in trade is able to pay only 17s. on a pound, what part of each pound will he pay? How much will he pay on a debt of 147£. 14s.?

62. What is $\frac{1}{3}$ of $\frac{21}{5}$?
63. Divide $\frac{24}{125}$ by 6.
64. Multiply $\frac{24}{125}$ by $\frac{1}{6}$.
65. What is $\frac{1}{15}$ of $\frac{4}{7}$?
66. Multiply $\frac{43}{47}$ by $\frac{1}{25}$.
67. Divide $\frac{43}{47}$ by 25.
68. Divide 15$\frac{33}{85}$ by 8.
69. Multiply 15$\frac{33}{85}$ by $\frac{1}{8}$.
70. What is $\frac{37}{285}$ of 17$\frac{2}{13}$?
71. Multiply 13$\frac{3}{7}$ by $\frac{4}{15}$.
72. Multiply 135$\frac{4}{11}$ by 24$\frac{2}{3}$.
73. Multiply 1,647$\frac{1}{4}$ by 17$\frac{13}{14}$.
74. How many times is 3 contained in 14$\frac{2}{3}$?
75. How many times is 9 contained in 47$\frac{4}{15}$?
76. How many times is 17 contained in 253$\frac{11}{17}$?
77. What part of 2 is $\frac{3}{7}$?
78. What part of 7 is $\frac{4}{15}$?
79. What part of 19 is $\frac{47}{48}$?
80. What part of 123 is $\frac{73}{185}$?
81. What part of 8 is 7$\frac{2}{3}$?
82. What part of 19 is 14$\frac{1}{7}$?
83. What part of 82 is 19$\frac{2}{15}$?
84. What part of 125 is 47$\frac{6}{15}$?

XVIII. 1. If 1 lb. butter cost $\frac{1}{4}$ of a dollar, how much will 2 lb. cost? What will 4 lb. cost?

2. At $\frac{1}{4}$ of a dollar per lb., what will 2 lb. of raisins cost? What will 3 lb. cost? What will 6 lb. cost?

3. If 1 man will consume $\frac{4}{5}$ of a bushel of corn in a week, how much will 2 men consume in the same time? How

much will 4 men consume? How much will 8 men consume?

4. If a horse will consume $\frac{4}{5}$ of a bushel of oats in a day, how much will he consume in 3 days? How much in 9 days?

5. If 1 man can do $\frac{1}{15}$ of a piece of work in a day, how much of it can 2 men do in the same time? How much of it can 3 men do? How much of it can 4 men do? How much of it can 6 men do? How much of it can 12 men do?

6. If a man drink $\frac{3}{40}$ of a barrel of cider in a week, how much would he drink in 2 weeks? How much would 5 men drink in a week at that rate? How much would 8 men drink in a week? How much would 20 men drink in a week? How much would 40 men drink in a week?

7. If a horse consume $2\frac{2}{8}$ bushels of oats in a week, how much would he consume in 4 weeks? How much in 8 weeks?

8. At $7\frac{3}{20}$ dollars a barrel, what cost 5 barrels of flour?

9. If a horse will eat $\frac{43}{248}$ of a ton of hay in a month, how much will 2 horses eat? How much will 8 horses eat?

10. If it take $1\frac{19}{24}$ yard of cloth to make a coat, how much will it take to make 8 coats? How much to make 24 coats?

11. If a barrel of cider cost $3\frac{27}{100}$ dollars, what will 10 barrels cost? What will 25 barrels cost?

12. Multiply $\frac{4}{25}$ by 5.
13. Multiply $\frac{43}{48}$ by 8.
14. Multiply $\frac{73}{125}$ by 25.
15. Multiply $\frac{37}{216}$ by 8.
16. Multiply $\frac{215}{841}$ by 9.
17. Multiply $\frac{217}{684}$ by 4.
18. Multiply $\frac{487}{1000}$ by 100.
19. Multiply $43\frac{27}{36}$ by 28.
20. Multiply $137\frac{19}{337}$ by 3.
21. Multiply $\frac{7}{8}$ by 8.

Note. 8 times $\frac{1}{8}=1$; 8 times $\frac{7}{8}$ is 7 times as much, that is, 7. Perform the following examples in a similar manner.

22. How much is 7 times $\frac{4}{7}$?
23. How much is 19 times $\frac{15}{19}$?
24. How much is 23 times $\frac{11}{23}$?
25. Multiply $7\frac{3}{4}$ by 5.
26. Multiply $19\frac{4}{17}$ by 17.
27. Multiply $123\frac{7}{9}$ by 9.
28. Multiply $43\frac{13}{327}$ by 327.

29. Multiply $9\frac{617}{1268}$ by 1268.

30. Multiply $14\frac{95}{1000}$ by 1000.

XIX. 1.* A merchant bought 4 pieces of cloth, the first contained $18\frac{3}{5}$ yards, the second $27\frac{1}{4}$ yards, the third $23\frac{1}{4}$ yards, and the fourth $25\frac{3}{4}$ yards. How many yards in the whole ? •

2. A gentleman hired 2 men and a boy for 1 week. One man was to receive $5\frac{3}{8}$ dollars, the other $7\frac{5}{8}$, and the boy $3\frac{7}{8}$. How much did he pay the whole ?

3. A gentleman hired three men for 1 month. To the first he paid $26\frac{3}{10}$ bushels of corn ; to the second, $28\frac{7}{10}$ bushels, and to the third, $33\frac{9}{10}$ bushels. How many bushels did it take to pay them ?

4. A man had $2\frac{1}{2}$ bushels of corn in one sack, and $2\frac{3}{4}$ in another ; how many bushels had he in both ?

5. If it takes $1\frac{1}{5}$ yard of cloth to make a coat, and $\frac{4}{5}$ of a yard to make a pair of pantaloons, how much will it take to make both ?

6. A man bought 2 boxes of butter ; one had $7\frac{3}{4}$ lb. in it, and the other $10\frac{3}{4}$ lb. How many pounds in both ?

7. A boy having a pine apple, gave $\frac{1}{4}$ of it to one sister, $\frac{1}{4}$ to another, and $\frac{1}{4}$ to his brother, and kept the rest himself. How much did he keep himself ?

8. A man bought 3 sheep ; for the first he gave $6\frac{3}{4}$ dollars ; for the second, $8\frac{4}{8}$; and for the third, $9\frac{1}{2}$. How many dollars did he give for the whole ?

9. How many cwt. of cotton in four bags containing as follows ; the first $4\frac{3}{4}$ cwt. ; the second, $5\frac{3}{7}$ cwt. ; the third $4\frac{9}{10}$ cwt ; and the fourth $7\frac{3}{20}$ cwt. ?

10. A merchant bought a piece of cloth containing 23 yards, and sold $7\frac{3}{4}$ yards of it ; how many yards had he left ?

11. A gentleman paid a man and a boy for 2 months' labour with corn ; to the man he gave $26\frac{2}{7}$ bushels, and to the boy he gave $18\frac{2}{3}$ bushels. How many bushels did it take to pay both ?

12. Bought $8\frac{2}{3}$ cwt. of sugar at one time, and $5\frac{2}{3}$ cwt. at another ; how much in the whole ?

13. Bought $\frac{3}{5}$ of a ton of iron at one time, and $\frac{4}{7}$ of a ton at another ; how much in the whole ?

14. There is a pole standing so that $\frac{3}{8}$ of it is in the mud,

* See First Lessons, Sect. XIII.

$\frac{2}{5}$ of it in the water, and the rest above the water; how much of it is above the water?

15. A merchant bought $14\frac{11}{13}$ cwt. of sugar, and sold $8\frac{5}{13}$ cwt.; how many lb. had he left?

Note. Reduce all fractions to their lowest terms, after the work is completed, or before if convenient. In the above example $\frac{5}{13}$ might be reduced, but it would not be convenient because it now has a common denominator with $\frac{11}{13}$. The answer may be reduced to lower terms.

16. Out of a barrel of cider there had leaked $7\frac{3}{4}$ gallons how many gallons were there left?

17. A man bought 2 loads of hay, one contained $17\frac{3}{4}$ cwt. and the other $23\frac{4}{13}$ cwt. How many cwt. in both?

18. A man had $43\frac{3}{4}$ cwt. of hay, and in 3 weeks his horse ate $5\frac{8}{17}$ cwt. of it; how much had he left?

19. Two boys talking of their ages, one said he was $9\frac{3}{4}$ years old; the other said he was $4\frac{5}{11}$ years older. What was the age of the second?

20. A lady being asked her age, said that her husband was $37\frac{4}{5}$ years old, and she was not so old as her husband by $8\frac{9}{13}$ years. What was her age?

21. A lady being asked how much older her husband was than herself, answered, that she could not tell exactly; but when she was married her husband was $28\frac{4}{17}$ years old, and she was $22\frac{2}{3}$. What was the difference of their ages?

22. Add together $\frac{2}{7}$ and $\frac{4}{13}$.

23. Add together $\frac{4}{5}$, $\frac{2}{7}$, and $\frac{3}{4}$.

24. Add together $\frac{2}{13}$ and $\frac{4}{17}$.

25. Add together $13\frac{4}{15}$ and $17\frac{3}{20}$.

26. Add together $137\frac{2}{3}$, $26\frac{9}{13}$, and $243\frac{3}{4}$.

27. What is the difference between $\frac{2}{3}$ and $\frac{3}{4}$?

28. What is the difference between $\frac{4}{15}$ and $\frac{16}{33}$?

29. What is the difference between $13\frac{2}{11}$ and $8\frac{5}{11}$?

30. What is the difference between $137\frac{3}{4}$ and $98\frac{3}{4}$?

31. Subtract $38\frac{4}{19}$ from $53\frac{3}{35}$.

32. Subtract $284\frac{38}{173}$ from $813\frac{31}{51}$.

XX. 1. A man bought 15 cows for $345. What was the average price?

Note. Find the price of 3 cows, and then of 1 cow.

2. A merchant bought 16 yards of cloth for $84.64; what was it a yard?

6 *

3. A merchant bought 18 barrels of flour for $ 114.66, and sold it so as to gain $1.00 a bbl. How much did he sell it for per barrel ?

4. 21 men are to share equally a prize of 8,530 dollars, how much will they have apiece ?

5. A merchant sold a hogshead of wine for 113 dollars. How much was it a gallon ?

6. A ship's crew of 30 men are to share a prize of 847 dollars ; how much will they receive apiece ?

7. A man has 1.857 lb. of tobacco, which he wishes to put into 42 boxes, an equal quantity in each box. How much must he put into each box ?

8. In 4,847 gallons of wine, how many hogsheads ?

9. At $48.00 a barrel, how many barrels of brandy may be bought for $687.43 ?

10. At $90 dollars a ton, how many tons of iron may be bought for 2,486 dollars ?

11. If 23.000 cwt. of iron cost $92,368.75, how much is it per lb. ?

12. Divide 784 by 28.

13. Divide 1,008 by 36.

14. Divide 1,728 by 72.

15. Divide 2,352 by 56.

16. Divide 183 by 15.

17. Divide 487 by 18.

18. Divide 1,243 by 25.

19. Divide 37,864 by 63.

20. Divide 19,748 by 112.

21. Divide 4,383 by 30.

22. Divide 6,487 by 50.

23. Divide 1,673 by 400.

24. Divide 13,748 by 7,000.

25. Divide 100,780 by 250.

26. Divide 406,013 by 4,700.

27. Divide 3,000,406 by 306,000.

28. Divide 450,387 by 36,000.

29. Divide 78,407,300 by 42,000.

30. Divide 15,008,406 by 480,000.

XXI. 1. Find the divisors of each of the following numbers, 15, 18, 20, 21, 24, 28, 42, 48, 64, 72, 88, 98.

2. Find the divisors of each of the following numbers, 108, 112, 114, 120, 387, 432, 846, 936.

3. Find the divisors of each of the following numbers, 8000, 4,053, 1,864, 2,480, 24,876, 103,284, and 7,328,472.

4. Find the common divisors of 8 and 24.

5. Find the common divisors of 16 and 36

6. Find the common divisors of 18 and 42

7. Find the common divisors of 21 and 56.

8. Find the common divisors of 56 and 264.

9. Find the common divisors of 123 and 642.

10. Find the common divisors of 32, 96, and 1,432.

11. Find the common divisors of 7,362, and 2,484.

12. Find the common divisors of 73,647, 84,177, and 9,684.

13. Reduce $\frac{14}{44}$ to its lowest terms.

14. Reduce $\frac{48}{300}$ to its lowest terms.

15. Reduce $\frac{300}{420}$ to its lowest terms.

16. Reduce $\frac{96}{480}$ to its lowest terms.

17. Reduce $\frac{486}{9720}$ to its lowest terms.

18. Reduce $\frac{4746}{38433}$ to its lowest terms.

19. Reduce $\frac{800}{42000}$ to its lowest terms.

XXII. 1. Reduce $\frac{3}{4}$ and $\frac{2}{5}$ to the least common denominator.

2. Reduce $\frac{3}{4}$ and $\frac{4}{18}$ to the least common denominator.

3. Reduce $\frac{2}{5}$ and $\frac{3}{8}$ to the least common denominator.

4. Reduce $\frac{3}{4}$ and $\frac{5}{14}$ to the least common denominator.

5. Reduce $\frac{5}{12}$ and $\frac{7}{18}$ to the least common denominator.

6. Find the least common multiple of 8 and 12.

7. Find the least common multiple of 8 and 14.

8. Find the least common multiple of 9 and 15.

9. Find the least common multiple of 15 and 18.

10. Find the least common muluple of 10, 14, and 15.

11. Find the least common multiple of 15, 24, and 35.

12. Find the least common multiple of 30, 48, and 56.

13. Find the least common multiple of 32, 72, and 120.

14. Find the least common multiple of 42, 60, and 125.

15. Find the least common multiple of 250, 180, and 540.

16. Reduce $\frac{3}{32}$ and $\frac{5}{28}$ to the least common denominator.

17. Reduce $\frac{4}{27}$ and $\frac{7}{54}$ to the least common denominator.

18. Reduce $\frac{5}{18}$, $\frac{8}{27}$, and $\frac{17}{90}$, to the least common denominator.

19. Reduce $\frac{2}{3}$, $\frac{5}{6}$, $\frac{7}{12}$, and $\frac{8}{27}$, to the least common denominator.

20. Reduce $\frac{4}{35}$, $\frac{3}{65}$, and $\frac{2}{15}$ to the least common denominator.

21. Reduce $\frac{13}{284}$ and $\frac{47}{648}$ to the least common denominator.

22. Reduce $\frac{35}{3500}$ and $\frac{43}{18000}$ to the least common denominator.

23. Reduce $\frac{115}{1250}$ and $\frac{840}{14400}$ to the least common denominator.

24. Reduce $\frac{174}{38800}$ and $\frac{38}{48000}$ to the least common denominator.

XXIII. 1.* At $\frac{1}{3}$ of a dollar a bushel, how many bushels of potatoes may be bought for 5 dollars ? How many at $\frac{2}{3}$ of a dollar a bushel ?

2. At $\frac{1}{5}$ of a shilling apiece, how many peaches may be bought for a dollar ? How many at $\frac{2}{5}$ of a shilling apiece ?

3. A gentleman distributed 6 bushels of corn among some labourers, giving them $\frac{1}{4}$ of a bushel apiece ; how many did he give it to ? How many would he have given it to, if he had given $\frac{3}{4}$ of a bushel apiece ?

4. If it takes $\frac{5}{6}$ of a bushel of rye to sow 1 acre, how many acres will 15 bushels sow ?

5. A merchant had 47 cwt. of tobacco which he wished to put into boxes, containing $\frac{7}{10}$ cwt. each. How many boxes must he get ?

6. A gentleman has a hogshead of wine which he wishes to put into bottles, containing $\frac{4}{15}$ of a gallon each. How many bottles will it take ?

7. If $\frac{3}{20}$ of a barrel of cider will last a family 1 week, how many weeks will 7 barrels last ?

8. If $\frac{7}{33}$ of a bushel of grain is sufficient for a family of two persons 1 day, how many days would 16 bushels last ? How many persons would 16 bushels last 1 day ?

9. If a labourer drink $\frac{13}{38}$ of a gallon of cider in a day, one day with another, how long will it take him to drink a hogshead ?

10. If an axe-maker put $\frac{9}{10}$ of a lb. of steel into an axe, how many axes would 1 cwt. of steel be sufficient for ?

11. If it take $1\frac{1}{2}$ bushel of oats to sow an acre, how many acres will 18 bushels sow ?

12. If it take $1\frac{1}{4}$ bushel of wheat to sow an acre, how many acres will 23 bushels sow ?

* See First Lessons, Sect. XV.

13. At $1\frac{3}{5}$ dollar a bushel, how much wheat may be bought for 20 dollars ?

14. At $3\frac{4}{7}$ dollars a barrel, how many barrels of cider may be bought for 40 dollars ?

15. At the rate of $15\frac{3}{4}$ bushels to the acre, how many acres will it take to produce 75 bushels of rye ?

16. At $4\frac{3}{8}$ dollars per cwt., how many tons of iron can I buy for $150 ?

17. At $11\frac{2}{3}$ cents per lb., how much steel can I buy for $50.00 ?

18. If a man can perform a journey in 580 hours, how many days will it take him to perform it if he travel $9\frac{3}{10}$ hours in a day ?

19. How many coats may be made of 187 yards of cloth if $3\frac{4}{17}$ yards make 1 coat ?

20. In 43 yards how many rods ?

21. In 87 yards how many rods ?

22. In 853 feet how many rods ?

23. In 2,473 feet how many furlongs ?

24. In 43,872 feet how many miles ?

25. If 1 bushel of apples cost $\frac{1}{4}$ of a dollar, how many bushels may be bought for $\frac{3}{4}$ of a dollar ?

26. At $\frac{1}{6}$ of a dollar a dozen, how many dozen of lemons may be bought for $\frac{4}{5}$ of a dollar ? How many dozen for $1\frac{2}{3}$ dollar ?

27. At $\frac{2}{7}$ of a dollar a dozen, how many dozen of oranges may be bought for $\frac{4}{5}$ of a dollar ? How many for $2\frac{1}{3}$ dollars ?

28. At $\frac{3}{5}$ of a dollar a bushel, how many bushels of apples may be bought for $\frac{7}{8}$ of a dollar ? How many for $5\frac{3}{8}$ dollars ?

29. At $\frac{1}{6}$ of a dollar per lb., how many pounds of figs may be bought for $\frac{2}{3}$ of a dollar ? How many pounds for $1\frac{1}{2}$ dollar ?

30. At $\frac{1}{3}$ of a dollar a bushel, how many bushels of apples may be bought for $1\frac{1}{4}$ dollar ?

31. If $\frac{1}{4}$ of a chaldron of coal will supply a fire 1 week, how many weeks will $\frac{3}{4}$ of a chaldron supply it ?

32. If 1 lb. of sugar cost $\frac{1}{8}$ of a dollar, how many pounds may be bought for $\frac{3}{4}$ of a dollar ? How many pounds for $1\frac{1}{3}$ dollar ?

33. At $\frac{1}{8}$ of a dollar per bushel, how many bushels of apples may be bought for $\frac{4}{7}$ of a dollar ? How many at $\frac{3}{8}$ of a dollar per bushel ?

34. At $\frac{1}{4}$ of a dollar per bushel, how many bushels of potatoes may be bought for $\frac{4}{5}$ of a dollar? How many at $\frac{2}{7}$ of a dollar per bushel?

35. At $\frac{3}{4}$ of a dollar a bushel, how much corn may be bought for $\frac{1}{4}$ of a dollar? How much for $\frac{1}{3}$ of a dollar?

36. At $\frac{5}{6}$ of a dollar per bushel, how much rye may be bought for $\frac{1}{3}$ of a dollar? How much for $\frac{3}{4}$ of a dollar?

37. At $\frac{1}{12}$ of a shilling apiece, how many eggs may be bought for $\frac{3}{4}$ of a dollar?

38. If it take $\frac{1}{12}$ of a pound of flour to make a penny-loaf, how many penny-loaves may be made of $\frac{3}{4}$ of a pound?

39. If a four-penny loaf weigh $\frac{4}{15}$ of a pound, how many will weigh $\frac{3}{4}$ of a pound?

40. If a two-penny loaf weigh $\frac{2}{15}$ of a pound, how many will weigh $1\frac{1}{4}$ lb? How many will weigh $7\frac{1}{4}$ lb?

41. If a six-penny loaf weigh $\frac{6}{16}$ of a pound, how many six-penny loaves will weigh $\frac{7}{8}$ of a pound? How many will weigh $4\frac{3}{8}$ lb?

42. If $\frac{5}{8}$ of a pound of fur is sufficient to make a hat, how many hats may be made of $4\frac{7}{16}$ lb. of fur?

43. If 10 oz. of fur is sufficient to make a hat, how many hats may be made of 4 lb. 7 oz. of fur?

44. If 1 bushel of apples cost $\frac{5}{9}$ of a dollar, how many bushels may be bought for $3\frac{4}{5}$ dollars?

45. If a bushel of apples cost 2s. 5d. how many bushels may be bought for 3 dollars and 5 shillings?

46. If $1\frac{3}{4}$, that is, $\frac{7}{4}$ of a yard of cloth will make a coat, how many coats may be made from a piece containing $43\frac{7}{8}$ yards?

47. If $2\frac{1}{4}$ bushels of oats will keep a horse 1 week, how long will $18\frac{5}{8}$ bushels keep him?

48. If $4\frac{3}{7}$ yards of cloth will make a suit of clothes, how many suits will $87\frac{4}{5}$ yards make?

49. If a man can build $4\frac{5}{7}$ rods of wall in a day, how many days will it take him to build $84\frac{4}{15}$ rods?

50. If $\frac{2}{4}$ of a ton of hay will keep a cow through the winter, how many cows will $23\frac{8}{15}$ tons keep at the same rate?

51. At $9\frac{24}{25}$ dollars a chaldron, how many chaldrons of coal may be bought for $37\frac{4}{5}$ dollars?

52. At $14\frac{7}{15}$ dollars per cwt., how many cwt. of yellow ochre may be bought for $243\frac{13}{17}$ dollars?

53. At $25\frac{3}{12}$ dollars a cask, how many casks of claret wine may be bought for $387\frac{5}{18}$ dollars?

54. At 95$\frac{15}{27}$ dollars a ton, how much iron may be bought for 2,956$\frac{1}{4}$ dollars ?

55. How many times is $\frac{5}{37}$ contained in 17 ?

56. How many times is $\frac{1}{8}\frac{3}{8}$ contained in 83 ?

57. How many times is 19$\frac{45}{57}$ contained in 253 ?

58. How many times is 42$\frac{25}{107}$ contained in 1,677 ?

59. How many times is $\frac{7}{8}$ contained in 14$\frac{1}{4}$?

60. How many times is $\frac{4}{15}$ contained in 37$\frac{5}{8}$?

61. How many times is 3$\frac{5}{7}$ contained in 24$\frac{4}{8}$?

62. How many times is 15$\frac{4}{17}$ contained in 103$\frac{13}{14}$?

63. How many times is 27$\frac{3}{37}$ contained in 1,605$\frac{5}{8}$.

64. At 3 dollars a barrel, what part of a barrel of cider may be bought for $\frac{1}{4}$ of a dollar ?

65. At 7 dollars a barrel, what part of a barrel of flour may be bought for $\frac{1}{3}$ of a dollar ? What part for $\frac{2}{3}$ of a dollar ?

66. At 11$\frac{1}{4}$ dollars per cwt., what part of 1 cwt. of sugar may be bought for $\frac{1}{4}$ of a dollar ? What part of 1 cwt. may be bought for $\frac{3}{4}$ of a dollar ? What part for 3$\frac{1}{4}$ dollars ?

67. At 93$\frac{1}{4}$ dollars per ton, what part of a ton of iron may be bought for 25$\frac{1}{8}$ dollars ?

68. When corn is $\frac{7}{8}$ of a dollar a bushel, what part of a bushel may be bought for $\frac{2}{3}$ of a dollar ?

69. Two men bought a barrel of flour, one gave 2$\frac{1}{4}$ dollars and the other 3$\frac{3}{4}$ dollars, what did they give for the whole barrel ? What part of the whole value did each pay ? What part of the flour should each have ?

70. Two men hired a pasture for 21 dollars. One kept his horse in it 5$\frac{1}{2}$ weeks, and the other 7$\frac{4}{8}$ weeks ; what ought each to pay ?

71. What part of 7$\frac{3}{4}$ is 2$\frac{4}{5}$?

72. What part of 53$\frac{3}{4}$ is 13$\frac{4}{5}$?

73. What part of 107$\frac{5}{18}$ is 93$\frac{4}{11}$?

74. What part of 3,840$\frac{3}{11}$ is $\frac{4}{37}$?

75. What part of $\frac{3}{4}$ is $\frac{3}{17}$?

76. What part of 11$\frac{3}{4}$ is 1$\frac{13}{14}$?

77. What part of 28$\frac{3}{15}$ is 13$\frac{1}{4}$?

78. What part of 137$\frac{3}{43}$ is 97$\frac{3}{37}$?

79. What part of 387$\frac{5}{37}$ is $\frac{2}{113}$?

XXIV. 1.* If $\frac{1}{2}$ of a gallon of brandy cost \$0.75, what is that a gallon ?

* See First Lessons, Sect. VI. and XI.

2. If $\frac{1}{2}$ of a ton of hay cost $13.375, what is that a ton ?

3. If $\frac{1}{3}$ of a yard of cloth cost $2.875 what is that a yard ?

4. If $\frac{1}{4}$ of a hhd. of brandy cost $27.00, what will 1 hhd. cost at that rate ?

5. A merchant bought $\frac{1}{2}$ of a pipe of brandy for $38.56 ; what would the whole pipe come to at that rate ?

6. A smith bought $\frac{1}{3}$ of a ton of iron for $12.43 ; what would a ton cost at that rate ?

7. A merchant owned $\frac{1}{13}$ of a ship's cargo, and his share was valued at $8,467.00 ; what was the whole ship valued at ?

8. A gentleman owned stock in a bank to the amount of $8,642.00, which was $\frac{1}{27}$ of the whole stock in the bank ; what was the whole stock ?

9. A gentleman lost at sea $4,843.67, which was $\frac{1}{38}$ of his whole estate ; how much was his whole property worth ?

10. A gentleman bought stock in a bank to the amount of $873.14, which was $\frac{1}{537}$ of the value of his whole property. What was the value of his whole property ?

11. A man bought $\frac{1}{2}$ of a bushel of corn for $\frac{1}{3}$ of a dollar ; what would be the price of a bushel at that rate ?

12. A man bought $\frac{1}{3}$ of a bushel of rye for $\frac{1}{4}$ of a dollar ; what would a bushel cost at that rate ?

13. A man sold $\frac{1}{5}$ of a yard of cloth for $\frac{2}{7}$ of a dollar ; what would a yard cost at that rate ?

14. A grocer sold $\frac{1}{8}$ of a gallon of wine for $\frac{3}{10}$ of a dollar ; what was it a gallon ?

15. A grocer sold $\frac{1}{37}$ of a barrel of flour for $\frac{6}{35}$ of a dollar ; what was it a barrel ?

16. A merchant sold $\frac{1}{5}$ of a ton of iron for $19\frac{4}{5}$ dollars ; how much was it a ton ?

17. A merchant sold $\frac{1}{10}$ of a hhd. of brandy for $11\frac{6}{17}$; how much was it per hhd. ?

18. A ship of war having taken a prize, the captain received $\frac{1}{17}$ of the prize money. His share amounted to $3,487$\frac{13}{405}$. What was the whole prize worth ?

19. If $\frac{2}{4}$ of a gallon of molasses cost 20 cents, what will $\frac{1}{4}$ cost. What will a gallon cost ? This question is the same as the following : If 2 quarts of molasses cost 20 cents, what is it a quart ? How much a gallon ?

20. If $\frac{3}{4}$ of a gallon, that is 3 quarts, of molasses cost 24 cents, what will $\frac{1}{4}$, that is 1 quart, cost ?

21. If $\frac{3}{4}$ of a yard of cloth cost 6 dollars, what cost $\frac{1}{4}$? What will a yard cost ?

22. If $\frac{3}{8}$ of a gallon, that is 3 pints, of wine cost 90 cents, what will $\frac{1}{8}$, that is 1 pint, cost ?   What will a gallon cost ?

23. If $\frac{5}{6}$ of a gallon of brandy cost 95 cents, what will $\frac{1}{6}$ cost ?   What will a gallon cost ?

24. If $\frac{3}{7}$ of a yard of broadcloth cost $6.00, what will $\frac{1}{7}$ cost ?   What will a yard cost ?

25. If $\frac{4}{7}$ of a box of lemons cost $2.40, what will $\frac{1}{7}$ cost ?   What will the whole box cost ?

26. If $\frac{4}{7}$ of a hhd. of molasses cost $16.00, what will the whole hogshead cost ?

27. A man travelled 12 miles in $\frac{3}{10}$ of a day ; how far did he travel in $\frac{1}{10}$ of a day ?   How far would he travel in a day at that rate ?

28. A man bought $\frac{4}{7}$ of a barrel of flour for $4.85, what would be the price of a barrel at that rate ?

29. A man being asked his age answered, that he was 24 years old when he was married, and that he had lived with his wife $\frac{4}{6}$ of his whole life.   What part of his whole age is 24 years ?   What was his age ?

30. A smith bought $\frac{5}{13}$ of a ton of Russia iron for $25.35, what would be the price of a ton at that rate ?

31. Bought $\frac{2}{3}$ of a yard of cloth for $5.00, what would be the price of a yard at that rate ?

32. If $\frac{3}{8}$ of a gallon of molasses, that is, 3 pints, cost 17 cents, what will $\frac{1}{8}$,(1 pint,)cost ?   What will a gallon cost ?

33. If $\frac{5}{16}$ of a pound of snuff, (5 ounces,) cost 14 cents, what cost $\frac{1}{16}$ lb., (1 ounce.) ?

34. If $\frac{4}{13}$ of a chaldron of coal cost $5, what cost $\frac{1}{13}$? What is that a chaldron ?

35. A man travelled 4 miles in $\frac{2}{5}$ of an hour ; how far would he travel in an hour at that rate ?

36. If $\frac{3}{13}$ of a ship's cargo is worth $14,000, what is the whole cargo worth ?

37. A owns $\frac{4}{15}$ of a coal mine, and his share is worth $3,500.   What is the whole mine worth ?

38. If $\frac{8}{135}$ of the stock in a bank is worth $63,275, what is the whole stock worth ?

39. If $1\frac{2}{3}$ yard of cloth is worth $11, what is a yard worth ?

40. If $2\frac{1}{3}$ bushels of corn is worth 13 shillings, what is a bushel worth ?

41. If $8_{\frac{9}{13}}$ bushels of wheat cost $15, what is it a bushel? What would 50 bushels cost at that rate?

42. A man sold $51_{\frac{8}{13}}$ cwt. of sugar for $587; what would be the price of $17_{\frac{2}{3}}$ cwt. at that rate?

43. If $\frac{3}{4}$ of 1 lb. of butter cost $\frac{2}{3}$ of a dollar, what will $\frac{1}{4}$ of 1 lb. cost? What will 1 lb. cost?

44. If $\frac{3}{4}$ of 1 lb. of raisins cost $\frac{2}{15}$ of a dollar, what will $\frac{1}{4}$ of 1 lb. cost? What will 1 lb. cost?

45. If $\frac{7}{8}$ of a bushel of corn cost $\frac{4}{9}$ of a dollar, what is that a bushel?

46. If $\frac{8}{15}$ of a barrel of flour will serve a family $1\frac{3}{7}$ of a month, how long will one barrel serve them? How long will 5 barrels serve them?

47. If $\frac{3}{7}$ of a yard of cloth cost $4\frac{2}{3}$ dollars, what is that a yard? What will $17\frac{3}{8}$ yards cost at that rate?

48. If $\frac{4}{15}$ of a hhd. of wine cost $30\frac{2}{3}$ dollars, what will be the price of a hhd. at that rate?

49. If $3\frac{1}{4}$ cwt. of iron cost $$14\frac{5}{8}$, what is that per cwt.?

50. If $7\frac{1}{2}$ lb. of butter cost $$1\frac{7}{11}$, what would be the price of $27\frac{1}{2}$ lb. at that rate?

51. A merchant bought a piece of cloth containing $24\frac{1}{2}$ yards, and in exchange gave $32\frac{3}{7}$ barrels of flour; how much flour did one yard of the cloth come to? How much cloth did 1 barrel of the flour come to?

52. If $\frac{5}{6}$ of a yard of cloth cost $\frac{3}{8}$ of a pound, what will $\frac{9}{11}$ of an ell English cost?

53. If $\frac{3}{4}$ of a barrel of flour cost $1\frac{3}{4}£$., what will $43\frac{1}{3}$ barrels cost?

54. A person having $\frac{2}{3}$ of a vessel, sells $\frac{3}{7}$ of his share for $8,400.00, what part of the whole vessel did he sell? What was the whole vessel worth?

55. If $\frac{1}{6}$ of a ship be worth $\frac{2}{7}$ of her cargo, the cargo being valued at $2,000£$., what is the whole ship and cargo worth?

56. If by travelling $12\frac{1}{2}$ hours in a day, a man perform a journey in $7\frac{3}{4}$ days, in how many days will he perform it, if he travel but $9\frac{1}{2}$ hours in a day?

57. If 5 men mow $72\frac{3}{4}$ acres in $11\frac{2}{3}$ days, in how many days will 8 men do the same?

58. If 5 men mow $72\frac{1}{2}$ acres in $11\frac{1}{4}$ days, how many acres will they mow in $8\frac{1}{4}$ days?

59. There is a pole, standing so that $\frac{2}{7}$ of it is in the water, $\frac{1}{4}$ as much in the mud as in the water, and $7\frac{3}{4}$ feet of it is above the water. What is the whole length of the pole?

60. A person having spent $\frac{1}{2}$ and $\frac{1}{4}$ of his money had $26\frac{2}{3}$ left. How much had he at first ?

61. Two men, A and B, having found a bag of money, disputed who should have it. A said $\frac{1}{2}$, $\frac{1}{3}$, and $\frac{1}{4}$ of the money made 130 dollars, and if B could tell him how much was in it he should have it all, otherwise, he should have nothing How much was in the bag ?

62. 45 is $\frac{5}{6}$ of what number ?

63. 486 is $\frac{9}{16}$ of what number ?

64. 68 is $\frac{5}{7}$ of what number ?

65. 125 is $\frac{13}{18}$ of what number ?

66. 376 is $\frac{23}{57}$ of what number ?

67. 17 is $\frac{43}{75}$ of what number ?

68. 3 is $\frac{29}{3083}$ of what number ?

69. 68 is $\frac{1}{987}$ of what number ?

70. 253 is $\frac{78}{1135}$ of what number ?

71. 37 is $\frac{125}{859}$ of what number ?

72. 6845 is $\frac{387}{1253}$ of what number ?

73. 384 is $\frac{1286}{13846}$ of what number ?

74. $\frac{3}{5}$ is $\frac{1}{4}$ of what number ?

75. $\frac{5}{6}$ is $\frac{3}{7}$ of what number ?

76. $\frac{1}{4}$ is $\frac{3}{8}$ of what number ?

77. $\frac{3}{14}$ is $\frac{4}{5}$ of what number ?

78. $\frac{11}{15}$ is $\frac{9}{13}$ of what number ?

79. $\frac{43}{28}$ is $\frac{6}{13}$ of what number ?

80. $\frac{74}{87}$ is $\frac{14}{37}$ of what number ?

81. $\frac{134}{567}$ is $\frac{4}{11}$ of what number ?

82. $\frac{16}{7}$ is $\frac{387}{423}$ of what number ?

83. $\frac{47}{15}$ is $\frac{135}{2387}$ of what number ?

84. $3\frac{5}{8}$ is $\frac{27}{16}$ of what number ?

85. $14\frac{3}{19}$ is $\frac{11}{135}$ of what number ?

86. $28\frac{4}{9}$ is $\frac{43}{328}$ of what number ?

87. $135\frac{11}{12}$ is $\frac{9}{16}$ of what number ?

88. $384\frac{5}{19}$ is $\frac{243}{87}$ of what number ?

89. $134\frac{3}{7}$ is $\frac{687}{123}$ of what number ?

90. Divide $134\frac{3}{7}$ by $\frac{687}{123}$.

91. $18\frac{53}{57}$ is $\frac{47}{11}$ of what number ?

92. Divide $18\frac{53}{57}$ by $\frac{47}{11}$.

93. $427\frac{3}{8}$ is $\frac{19}{4}$ of what number ?

94. Divide $42\frac{3}{8}$ by $2\frac{1}{4}$, that is $\frac{19}{4}$.

95. $384\frac{4}{13}$ is $\frac{11}{3}$ of what number ?

96. Divide $384\frac{4}{13}$ by $3\frac{2}{3}$ or $\frac{11}{3}$.

97. 42 is $\frac{5}{8}$ of what number ?

98. How many times is $\frac{4}{5}$ contained in 42 ?

99. Divide 42 by $\frac{4}{5}$.

100. $3\frac{4}{15}$ is $\frac{5}{7}$ of what number ?

101. How many times is $\frac{5}{7}$ contained in $3\frac{4}{15}$?

102. Divide $3\frac{4}{15}$ by $\frac{5}{7}$.

103. $13\frac{2}{7}$ is $\frac{16}{7}$ of what number ?

104. How many times is $2\frac{2}{7}$ or $\frac{16}{7}$ contained in $13\frac{2}{7}$?

105. Divide $13\frac{2}{7}$ by $2\frac{2}{7}$.

106. A merchant sold a quantity of goods for $252.00, which was $\frac{5}{6}$ of what it cost him ? How much did it cost him, and how much did he gain ?

107. A merchant sold a quantity of goods for $243.00, by which he gained $\frac{1}{8}$ of the first cost. What was the first cost, and how much did he gain ?

Note. If he gained $\frac{1}{8}$ of the first cost, $243.00 must be $\frac{9}{8}$ of the first cost.

108. A merchant sold a quantity of goods for $3,846.00, by which bargain he gained $\frac{1}{5}$ of the first cost. What was the first cost, and how much did he gain ?

109. A merchant sold a hhd. of wine for $108.43, by which bargain he gained $\frac{1}{7}$ of the first cost. What was the first cost per gallon ?

110. A merchant sold a bale of cloth for $347.00, by which he gained $\frac{3}{10}$ of what it cost him ? How much did it cost him, and how much did he gain ?

Note. If he gained $\frac{3}{10}$ of the first cost, $347.00 must be $1\frac{3}{10}$ of the first cost.

111. A merchant sold a quantity of flour for $147.00, by which he gained $\frac{2}{5}$ of the cost. How much did it cost, and how much did he gain ?

112. A merchant sold a quantity of goods for $6,487.00, by which he gained $\frac{7}{18}$ of the cost. How much did he gain?

113. A merchant sold a quantity of goods for $187.00 by which he lost $\frac{1}{4}$ of the first cost. How much did it cost, and how much did he lose ?

Note. If he lost $\frac{1}{4}$ of the cost, $187.00 must be $\frac{3}{4}$ of the cost.

114. A merchant sold a quantity of molasses for $258.00, by which he lost $\frac{1}{8}$ of the cost. How much did it cost, and how much did he lose ?

115. A merchant sold a quantity of goods for $948.00, by which he lost $\frac{4}{15}$ of the cost. How much did he lose ?

116. A merchant sold 3 hhds. of molasses for $67.23, by which he lost $\frac{2}{10}$ of the first cost. How much did he lose ? How much on a gallon ?

117. A merchant sold 93 yards of cloth for $527.43, by which he lost $\frac{2}{15}$ of the cost. How much did he lose on a yard ?

118. A merchant sold a quantity of goods so as to gain $43, which was $\frac{2}{7}$ of what the goods cost him. How much did they cost ?

119. A merchant sold a quantity of goods for $273.00, by which he gained 10 per cent. on the first cost. How much did they cost ?

Note. 10 per cent. is 10 dollars on a 100 dollars, that is, $\frac{10}{100}$. 10 per cent. of the first cost therefore is $\frac{10}{100}$ of the first cost. Consequently $273.00 must be $\frac{110}{100}$ of the first cost.

120. A merchant sold a quantity of goods for $135.00, by which he gained 13 per cent. How much did the goods cost, and how much did he gain ?

121. A merchant sold a quantity of goods for $3,875 by which he gained 65 per cent. How many dollars did he gain ?

122. A merchant sold a quantity of goods for $983.00, by which he lost 12 per cent. How much did the goods cost and how much did he lose ?

Note. If he lost 12 per cent., that is $\frac{12}{100}$, he must have sold it for $\frac{88}{100}$ of what it cost him.

123. A merchant sold 3 hhds. of brandy for $248.37, by which he lost 25 per cent. How much did the brandy cost him, and how much did he lose ?

124. A merchant sold a quantity of goods for $87.00 more than he gave for them, by which he gained 13 per cent. of the first cost. What did the goods cost him, and how much did he sell them for ?

Note. Since 13 per cent. is $\frac{13}{100}$, $87 must be $\frac{13}{100}$ of the first cost.

125. A merchant sold a quantity of goods for $43.00 more than they cost, and by doing so gained 20 per cent. How much did the goods cost him ?

7 *

126. A merchant sold a quantity of goods for $137.00 less than they cost him, and by doing so lost 23 per cent. How much did the goods cost, and how much did he sell them for?

127. A has tea which he sells B for 10d. per lb. more than it cost him, and in return B sells A cambrick, which cost him 10s. per yd., for 12s. 6d. per yard. The gain on each was in the same proportion. What did A's tea cost him per lb.?

Note. B gains 2s. 6d. on a yard, which is $\frac{1}{4}$ of the first cost, consequently 10d. must be $\frac{1}{4}$ of the first cost of the tea?

128. C has brandy which he sells to D for 20 cents per gal. more than it cost him; and D sells C molasses which cost 23 cents per gal. for 32 cents per gal., by which D gains in the same proportion as C. How much did C's brandy cost him per gal.?

129. A man being asked his age, answered, that if to his age $\frac{1}{2}$ and $\frac{1}{4}$ of his age be added, the sum would be 121. What was his age?

130. A man having put a sum of money at interest at 6 per cent., at the end of 1 year received 13 dollars for interest. What was the principal?

Note. Since 6 per cent. is $\frac{6}{100}$ of the whole, 13 dollars must be $\frac{6}{100}$ of the principal.

131. What sum of money put at interest for 1 year will gain 57 dollars, at 6 per cent.?

132. A man put a sum of money at interest for 1 year, at 6 per cent., and at the end of the year he received for principal and interest 237 dollars. What was the principal?

Note. Since 6 per cent. is $\frac{6}{100}$, if this be added to the principal it will make $\frac{106}{100}$, therefore $237 must be $\frac{106}{100}$ of the principal. When the interest is added to the principal the whole is called the *amount.*

133. What sum of money put at interest at 6 per cent. will gain $53 in 2 years?

Note. 6 per cent. for 1 year will be 12 per cent. for 2 years, 3 per cent. for 6 months, 1 per cent. for 2 months, &c.

134. What sum of money put at interest at 6 per cent will gain $97 in one year and 6 months?

135. What sum of money put at interest at 6 per cent. will amount to $394 in 1 year and 8 months?

136. What sum of money put at interest at 7 per cent. will amount to £183 in 1 year?

137. What sum of money put at interest at 8 per cent. will amount to $137 in 2 years and 6 months?

138. Suppose I owe a man $287 to be paid in one year without interest, and I wish to pay it now; how much ought I to pay him, when the usual rate is 6 per cent.?

Note. It is evident that I ought to pay him such a sum, as put at interest for 1 year will amount to $287. The question therefore is like those above. This is sometimes called *discount.*

139. A man owes $847 to be paid in 6 months without interest, what ought he to pay if he pays the debt now, allowing money to be worth 6 per cent. a year?

140. A merchant being in want of money sells a note of $100, payable in 8 months without interest. How much ready money ought he to receive, when the yearly interest of money is 6 per cent.?

141. According to the above principle, what is the difference between the interest of $100 for 1 year, at 6 per cent. and the discount of it for the same time?

142. What is the difference between the interest of $500 for 4 years at 6 per cent., and the discount of the same sum for the same time?

Miscellaneous Examples.

In measuring surfaces, such as land, &c. a square is used as the measure or unit. A square is a figure with four equal sides, and the four corners or angles equal. The square is used because it is more convenient for a measure than a figure of any other form. The figure A B C D is a square. The sides are each one inch, consequently it is called a square inch. A figure one foot long and one foot wide is called a square foot; a figure one yard long and one yard wide is called a square yard, &c.

1. If a figure one inch long and one inch wide contains one square inch, how many square inches does a figure one inch wide and two inches long contain? How many square inches does a figure one inch wide and three inches long contain? Four inches long? Five inches long? Seven inches long?

2. In a figure 8 inches long and 1 inch wide, how many square inches? How many square inches does a figure 8 inches long and 2 inches wide contain? 3 inches wide? 4 inches wide? 5 inches wide? 8 inches wide?

3. If a figure 1 foot wide and 1 foot long contains 1 square foot, how many square feet does a figure 1 foot wide and 2 feet long contain? How many square feet does a figure 1 foot wide and 3 feet long contain? 5 feet long? 9 feet long? 15 feet long?

4. In a figure 9 feet long and 1 foot wide, how many square feet? How many square feet does a figure 9 feet long and 2 feet wide contain? 3 feet wide? 5 feet wide? 7 feet wide? 9 feet wide?

5. How many square inches does a figure 13 inches long and 1 inch wide contain? 2 inches wide? 3 inches wide? 8 inches wide?

6. How many square feet does a figure 16 feet long and 1 foot wide contain? 2 feet wide? 3 feet wide? 5 feet wide? 8 feet wide? 13 feet wide?

In the above examples supply yards, rods, furlongs, and miles, instead of inches and feet, and perform them again.

7. What rule can you make for finding the number of square inches, feet, yards, &c. in any rectangular figure?

Note. A figure with four sides, which has all its angles alike or right angles, is called a *rectangle*, and a rectangle is called a *square* when all the sides are equal.

8. How many square feet in a room 18 feet long and 13 feet wide?

9. How many square feet in a piece of land 143 feet long and 97 feet wide?

10. How many square rods in a piece of land 28 rods long and 7 rods wide?

11. A piece of land that is 20 rods long and 8 rods wide, or in any other form containing the same surface, is called an acre. How many square rods in an acre?

12. How wide must a piece of land be that is 17 rods long to make an acre ?

13. How many square inches in a square foot ; that is, in a figure that is 12 inches long and 12 wide ?

14. How much in length, that is 8 inches wide, will make a square foot ?

15. How many square feet in a square yard ?

16. How many square yards in a square rod ?

17. How many square inches in a square yard ?

18. A piece of land 20 rods long and 2 rods wide, or in any other form which contains the same surface, is called a rood. How many square rods in a rood ?

19. How many roods make an acre ?

20. Find the numbers for the following table.

SQUARE MEASURE.

square inches	make	1 square foot
square feet		1 square yard
square yards or ⎰		1 square rod,
square feet ⎱		perch, or pole
square rods		1 rood
roods		1 acre

21. How many square inches in a square rod ?

22. How many square yards in an acre ?

23. How many square inches in an acre ?

24. How many square feet in 1728 square inches ?

25. In 286 square poles how many acres ?

26. In 201,283,876 square inches, how many acres ?

27. How many square rods in a square mile ?

28. How many acres in a square miles ?   ·

29. The whole surface of the globe is estimated at about 198,000,000 square miles. How many acres on the surface of the globe ?

30. How many square inches in a board 15 inches wide and 11 feet long ? How many square feet ?

31. How many acres in a piece of land 183 rods long and 97 rods wide ?

32. How many square inches in a yard of carpeting that is 2 ft. 3 in. wide ? How many yards of such carpeting will it take to cover a floor 19 ft. 4 in. long and 17 ft. 2 in. wide ?

To measure solid bodies, such as timber, wood, &c., it is necessary to use a measure that has three dimensions, length, breadth, and depth, height, or thickness. For this a measure is used in which all these dimensions are alike. Take a block, for example, and make it an inch long, an inch wide, and an inch thick, and all its corners or angles alike ; this is called a *solid* or *cubic* inch ; so a block made in the same way having each of its dimensions one foot, is called a *solid* or *cubic* foot.

33. If a block 1 inch wide and 1 inch thick and 1 inch long contains 1 solid inch, how many solid inches does such a block that is 2 inches long contain? 3 inches long? 4 inches long? 5 inches long? 8 inches long?

34. How many solid inches does a block that is 1 foot long, 1 inch thick, and 1 inch wide contain? How many inches does such a block that is 2 inches wide contain? 3 inches wide? 4 inches wide? 5 inches wide? 8 inches wide?

35. How many solid inches does a block 2 inches long, 2 inches wide, and 1 inch thick contain? 2 inches thick?

36. How many solid inches does a block 4 inches long, 3 inches wide, and 1 inch thick contain? 2 inches thick? 3 inches thick?

37. How many cubic inches in a block 10 inches long, 8 inches wide, and 1 inch thick? 2 inches thick? 3 inches thick? 5 inches thick? 7 inches thick?

38. How many cubic inches in a block 18 inches long, 13 inches wide, and 1 inch thick? 5 inches thick? 11 inches thick?

In the above examples supply feet instead of inches, and do them over again.

39. What rule can you make for finding the number of solid inches or feet in any regular solid body?

40. How many solid inches in a block 12 inches long, 12 inches wide, and 12 inches thick; that is, in a solid foot?

41. A pile of wood 8 feet long, 4 feet wide, and 4 feet high, or in any other form containing an equal quantity, is called a *cord* of wood. How many solid feet in a cord?

42. Find the numbers for the following table.

SOLID OR CUBIC MEASURE.

	solid inches	make	1 solid foot
	solid feet		1 cord of wood
40	solid feet of round timber, or ⎱		
50	solid feet of hewn timber ⎰		1 ton or load

43. How many solid inches in a cord?

44. How many solid inches in a ton of hewn timber?

45. In 468,374 solid inches, how many solid feet?

46. How many feet of timber in a stick 28 feet long and 11 inches square?

47. How many tons of timber in 2 sticks, each 25 feet long, 15 inches wide, and 11 inches thick?

48. A pile of wood 4 feet square and 1 foot long, or a pile containing 16 solid feet is called 1 *foot of wood*. How many such feet in a cord?

49. How many solid feet of wood in a pile 5 feet wide, 3 feet high, and 23 feet long? How many feet of wood? How many cords?

A few more examples of this kind will be found in decimals.

DECIMAL FRACTIONS.

XXV. In the following numbers, write the fractional part in the form of decimals.

1. Twenty-seven and six tenths, $27\frac{6}{10}$.   *Ans.* 27.6.

2. Fourteen and seven hundredths. $14\frac{7}{100}$.

          Ans. 14.07.

3. One hundred twenty-three, and eight thousandths. $123\frac{8}{1000}$.   *Ans.* 123.008.

4. One hundred and eight, and five tenths. $108\frac{5}{10}$.

5. Seventy-three, and nine hundredths. $73\frac{9}{100}$.

6. Four, and six thousandths. $4\frac{6}{1000}$.

7. Sixteen, and one thousandth. $16\frac{1}{1000}$.

8. Six tenths. $\frac{6}{10}$.

9. Five hundredths. $\frac{5}{100}$.

10. Seven thousandths. $\frac{7}{1000}$.

11. Two ten thousandths. $\frac{2}{10000}$.

12. Three, and four tenths and two hundredths. $3\frac{4}{10}$ and $1\frac{2}{100}$.

13. $\frac{4}{10}$ are how many hundredths ?

14. $\frac{4}{10}$ and $\frac{2}{100}$ are how many hundredths ?

15. $\frac{3}{10}$ are how many thousandths ?

16. $\frac{8}{100}$ are how many thousandths ?

17. $\frac{3}{10}$ and $\frac{8}{100}$ and $\frac{5}{1000}$ are how many thousandths ?

18. Write $7\frac{385}{1000}$ in the form of a decimal.

19. $\frac{2}{10}$ are how many ten-thousandths ?

20. $\frac{5}{100}$ are how many ten-thousandths ?

21. $\frac{6}{1000}$ are how many ten-thousandths ?

22. $\frac{2}{10}$, $\frac{5}{100}$, $\frac{6}{1000}$, and $\frac{7}{10000}$ are how many ten-thousandths ?

23. Write $\frac{2567}{10000}$ in the form of a decimal ?

Write the fractions in the following numbers in the form of decimals.

24. $13\frac{23}{100}$.

25. $21\frac{182}{1000}$.

26. $12\frac{5736}{10000}$.

27. $142\frac{38746}{100000}$.

28. $1\frac{43}{1000}$.

29. $17\frac{573}{10000}$.

30. $193\frac{47}{10000}$.

31. $87\frac{106}{100000}$.

32. $95\frac{406}{1000}$.

33. $98\frac{6004}{1000000}$.

34. $\frac{30507}{100000}$.

35. $\frac{807}{10000}$.

Change the decimals in the following numbers to common fractions and reduce them to their lowest terms.

36. 42.5.

37. 84.25.

38. 9.8.

39. 137.16.

40. 25.125.

41. 18.625.

42. 11.8642.

43. 163.90064. •

44. 72.0065.

45. 4.00025.

46. 13.0060058.

47. 0.75.

48. 0.3125.

49. .075.

50. .00128.

51. .00015.

52. .000106.

53. .1500685.

XXVI. 1. A man purchased a barrel of flour for $7.43.; 5 gallons of molasses for $1.625; 3 gallons of wine for $4.87; 4 gallons of brandy for $7; 7 lbs. of sugar for $0.95; and 3 gallons of vinegar for $0.42. What did the whole amount to ?

2. How many bushels of corn in 4 bags, the first containing $2\frac{8}{10}$ bushels; the second, $3\frac{28}{100}$; the third, $3\frac{42}{1000}$; and the fourth, $4\frac{287}{1000}$?

Note. Write the fractions in the form of decimals.

3. A man bought four loads of hay, the first containing $17\frac{3}{5}$ cwt. ; the second, $19\frac{1}{4}$ cwt. ; the third, $24\frac{4}{5}$ cwt. ; and the fourth, $14\frac{1}{2}$ cwt. How many cwt. in the whole ?

Note. In all the examples under the head of decimals, change the fractions and parts to decimals.

4. A man raised wheat in five fields, in the first, $47\frac{3}{10}$ bushels ; in the second, $94\frac{8}{25}$? in the third, $87\frac{14}{25}$; in the fourth, $143\frac{11}{16}$; and in the fifth 387 bushels. How many bushels in the whole ?

5. A man bought a load of hay for $6\frac{4}{25}£.$; a load of oats for $7\frac{7}{40}£.$; 3 bushels of corn for $\frac{17}{20}£.$; and a load of wood for $2\frac{9}{50}£.$ How much did the whole come to ?

6. Add together the following numbers, $38\frac{1}{4}$; $1386\frac{7}{10}$; 7006 ; $\frac{97}{250}$; $\frac{400}{2875}$; 8 ; and $460\frac{13}{88}$.

7. From a piece of cloth containing $47\frac{3}{8}$ yards, a merchant sold $23\frac{9}{25}$. How much remained unsold ?

8. A man owing \$253 paid \$187.375, how much did he then owe ?

9. A man owing $342\frac{64}{125}£.$ paid $187\frac{4}{13}£.$ How much did he then owe ?

10. A merchant sold a barrel of flour for $2\frac{5}{12}£.$; 5 gallons of molasses for $\frac{11}{16}£.$; and 6 gallons of wine for $2\frac{9}{43}£.$ In pay he received a load of wood worth $2\frac{4}{15}£.$ and 2 bushels of wheat, worth $\frac{13}{14}£.$ and the rest in money ; how much money did he receive ?

11. From $183\frac{3}{4}£.$ take $87\frac{1}{5}£.$

12. From \$382 take \$48.25.

13. From $1153\frac{3}{7}$ lb. take $684\frac{5}{11}$ lb.

14. From $37\frac{2}{3}$ tons 'ake $28\frac{9}{17}$ tons.

Multiplication of Decimals.

XXVII. 1. A man bought 5 barrels of pork, at \$17.43 per barrel ; how much did it come to ?

2. What cost 8 yards of cloth, at \$7.875 per yard ?

3. How many bushels of meal in 14 sacks, containing 4.37 bushels each ?

4. How much hay in 8 loads, containing 24.35 cwt. each ?

5. How much cotton in 17 bales, containing $4\frac{3}{8}$ cwt. each ?

6. How many cwt. of hay in 14 loads, containing 23.25 cwt. each ?

7. Multiply 42.62 by 38.
8. Multiply 137.583 by 17.
9. Multiply 13.946 by 58.
10. Multiply 2.5837 by 15.
11. Multiply .464 by 27.
12. Multiply .0038 by 9.
13. If a barrel of flour cost $5, what cost .6 of a barrel ?
14. At $90 per hhd., what cost .7 hhd., that is, $\frac{7}{10}$ of a hdd. ?
15. At $45 per hhd., what cost .8 hhd., that is, $\frac{8}{10}$ of a hhd. of gin ?
16. At $20 per hhd., what cost 2.9 hhds., that is $2\frac{9}{10}$ hhds. of molasses ?
17. At $25 per ton, what cost 7.6 tons of hay ?
18. At $95 per ton, what cost 3.7 tons of iron ?
19. At $32 per ton, what cost 14.25 tons of logwood ?
20. At $220 per ton, what cost 19.47 tons of hemp ?
21. At $57 per ton, what cost 3.5 tons of alum ?
22. At $45 per thousand, what cost 2.5 thousands of staves ?
23. What is .5 of 128 ?
24. What is .25 of 856 ?
25. What is .125 of 856 ?
26. What is .287 of 2487 ?
27. Multiply 2487 by .287.
28. Multiply 4306 by 3.5.
29. Multiply 87 by 2.8.
30. Multiply 1864 by 3.25.
31. Multiply 30067 by 1.3873.
32. Multiply 10372 by $6\frac{1}{2}$=6.5.
33. Multiply 468 by $7\frac{1}{4}$=7.25.
34. Multiply 46800 by $13\frac{3}{4}$.
35. Multiply 36038 by $1\frac{3}{8}$.
36. Multiply 130407 by $5\frac{3}{35}$.
37. At .3 of a dollar a gallon, what cost .2 of a gallon of molasses ?
38. What is .2 of .3, that is $\frac{2}{10}$ of $\frac{3}{10}$?
39. Multiply .3 by .2.
40. At $.90 per gallon, what cost .4 of a gal. of wine ?
41. At $.25 per lb. what cost 2.8 lb. of butter ?
42. At $.36 per lb., what cost 4.5 lb. of sperm candles ?
43. At $.47 per piece, what cost 4.3 pieces of nankin ?
44. At $5.37 per yard, what cost 7.4 yards of cloth ?

45. At $13.50 per bbl., what cost 14¾ bbls. of pork ?
46. At $25.45 per ton, what cost 18⅝ tons of hay ?
47. At $140.50 per ton, what cost 13¼ tons of potashes ?
48. If an orange is worth $.06, what is .3 of an orange worth ?
49. If a bale of cotton contains 4.37 cwt., what is .45 of a bale ?
50. Multiply 4.5 by 2.3.
51. Multiply 13.43 by 1.4.
52. Multiply 43.25 by .8.
53. Multiply 284.43 by 1.02.
54. Multiply 18.325 by 1.38.
55. Multiply 6.4864 by 2.03.
56. Multiply 14.00643 by .5.
57. Multiply 3.400702 by 1.003.
58. Multiply 1.006 by .002.
59. Multiply 1.0007 by .0003.
60. Multiply .3 by .2.
61. Multiply .04 by .2.
62. Multiply .003 by .01.
63. Multiply .0004 by .025.
64. Multiply .0107 by .00103.
65. Multiply 1.340068 by 1.003084

Miscellaneous Examples.

1. At $12 per cwt. what cost 5 cwt. 3 qrs. of sugar ?

Note.   5 cwt. 3 qrs. is 5¾ cwt., that is 5.75 cwt.

2. At $25 per cwt., what cost 37 cwt. 3 qrs. 14 lb. of tobacco ?

Note.   The quarters and pounds may first be reduced to a common fraction and then to decimals.   3 qrs. 14 lb. are 98 lb., that is $\frac{98}{112}$ of 1 cwt., and $\frac{98}{112} = .875$; therefore, 37 cwt. 3 qrs. 14 lb. is equal to 37.875 cwt. ; this multiplied by 25 gives $946.875.

3. What cost 5 cwt. 2 qrs. 19 lb. of raisins, at $11 per cwt. ?

4. What cost 13 cwt. 1 qr. 15 lb. of iron, at $4.27 per cwt. ?

Note. 13 cwt. 1 qr. 15 lb.$=13\frac{4\cdot3}{17\cdot2}$ cwt.$=13.383+$cwt. This multiplied by $4.27 gives $57.14541. Observe, that there must be as many decimal places in the product as in the multiplicand and multiplier together. In this instance there are five places. It is not necessary to notice any thing smaller than mills in the result, therefore $57.145 will be sufficiently exact for the answer.

5. What cost 12 cwt. 0 qrs. 19 lb. of rice, at $3.28 per cwt. ?

6. What cost 13 cwt. 2 qrs. 4 lb. of hops, at $5.75 per cwt.

7. What cost 3 hhds. 43 gal. of wine, at $98 per hhd. ?

Note. 3 hhds. 43 gal. is $3\frac{43}{63}$ hhds. ; this reduced to a decimal is 3.683 hhds., nearly.

8. What cost 17 hhds. 18 gal. of molasses, at $23.25 per hhd. ?

9. What cost 13 hhds. 53 gal. of gin, at $47.375 per hhd. ?

10. What cost 4 hhds. 27 gal. 3 qts. of brandy, at $108.42 per hhd. ?

11. Express in decimals of an cwt. the quarters, pounds, and ounces in the following numbers :—3 cwt. 2 qrs. 22 lb. ; 17 cwt. 1 qr. 11 lb. 5 oz. ; 4 cwt. 0 qr. 16 lb. 3 oz.

12. Express in decimals of a hogshead the gallons, quarts, pints, &c. in the following numbers :—43 hhds. 17 gal. 2 qts ; 14 gal. 6 qts. 1 pt. ; 7 hhds. 0 gal. 3 qts. 1 pt.

13. What cost 8 gal. 3 qts. 1 pt. of gin, at $0.43 per gal. ?

14. What cost 17 lb. 13 oz. of sugar, at $0.12 per lb. ?

15. What cost 23lb. 7 oz. of sugar, at $11.43 per cwt. ?

16. What cost 11 gals. 2 qts. of brandy, at the rate of $98.48 per hhd. ?

17. What cost 17 yds. 3 qrs. 2 nls. of broadcloth, at $7.25 per yard ?

18. What cost 2 qrs. 3 nls. of broadcloth, at $6.42 per yard ?

Express the fractions in the following examples in decimals.

19. What part of 1 yd. is 3 qrs. 2 nls. ?

20. What part of 1 yard is 1 qr. 3 nls. ?

21. What part of 1 lb. Avoirdupois is 13 oz. ?

22. What part of 1 qr. is 17 lb. ?

23. What part of 1 qr. is 13 lb. 5 oz. ?

24. What part of a day is 6 hours ?
25. What part of a day is 16 h. 25 min. ?
26. What part of a day is 13 h. 42 min. 11 sec. ?
27. What part of an hour is 47 min. ?
28. What part of an hour is 38 min. 47 sec. ?
29. What part of a rod is 13 ft. ?
30. What part of 1 ft. is 2 in. ?
31 What part of 1 ft. is 7 in. ?
32. What part of a rod is 7 ft. 4 in. ?
33. What part of a mile is 7 rods, 13 ft. ?
34. What part of 1£. is 13s. 6d. ?
35. What part of 1s. is 5d. 1 qr ?
36. What part of 1£. is 11s. 5d. 3 qr. ?
37. At 2£. 5s. per cwt., what cost 5 cwt. 3 qrs. of raisins ?

Note. 2£. 5s.=2.25£., and 5 cwt. 3 qrs.=5.75 cwt. Multiplying these together, the result is 12.9375£. The decimal part of this result may be changed to shillings and pence again. .9375£. is .9375 of 20 shillings ; therefore if we multiply 20 shillings by .9375, or, which is the same thing, if we multiply .9375 by 20, we shall obtain the answer in shillings and parts of a shilling. This is evident also from another course of reasoning. .9375£. is now in pounds; if it be multiplied by 20 it will be reduced to shillings.

.9375
20
—————
18.7500   The result is 18 shillings and .75 of a shilling, which may in like manner be reduced to pence by multiplying it by 12.

.75
12
—————
9.00   The result is 9d. The answer, therefore, is 12£. 18s. 9d.

38. What cost 3 cwt. 2 qrs. 7 lb. of hops, at 2£. 3s. 6d. per cwt. ?
39. What cost 17 yds. 2 qrs. 2 nls. of broadcloth, at 2£. 5s. 7d. per yard ?
40. What cost 8 cwt. 1 qr. 13 lb. of wool, at 3£. 7s. 6d. per cwt. ?
41. What cost 3 hhds. 43 gals. of wine, at 32£. 14s. 8d. per hhd. ?

42. How many cwt. of raisins in 7⅗ casks, each cask containing 2 cwt. 0 qrs. 25 lbs?

Note. 7⅗=7.6 and 2 cwt. 0. qrs. 25 lb.=2.2232+ cwt. These multiplied together produce 16.8957 cwt. The fractional part of this may be changed to quarters, pounds, &c. as the fractions in the last examples were changed to shillings and pence. .8957 cwt. is .8957 of 4 quarters, or it is hundred-weights and may be reduced to quarters and pounds by multiplying by 4, and by 28.

```
      .8957
          4
      ------
      3.5828
         28
      ------
      46624
      11656
      ------
      16.3184
          16
      ------
       19104
        3184
      ------
       5.0944
```

The result is 3 qrs. and a fraction. Then multiply .5828 qrs. by 28, it gives 16 lb. and a fraction of a pound. Multiplying .3184 lb. by 16, it gives 5 oz. and a fraction of an ounce.

The answer is 16 cwt. 3 qrs. 16 lb. 5 1/16 oz. nearly. The same result may be obtained by changing the decimal .8957 cwt. to a common fraction, and proceeding according to the method given in Art. XVI.

43. How many cwt. of cotton in 5¾ bales, each bale containing 4 cwt. 3 qrs. 7 lb.?

44. How many cwt. of coffee in 13¾ bags, each bag containing 1 cwt. 3 qrs. 15 lb.?

45. Find the value of .387£. in shillings, pence, and farthings.

46. Find the value of .9842£. in shillings, pence, and farthings.

47. Find the value of .583 cwt. in quarters, pounds, &c.

48. Find the value of .23 cwt. in quarters, pounds, &c.

49. Find the value of .73648 cwt. in quarters, pounds, &c.

50. Find the value of .764s. in pence and farthings.

51. Find the value of .3846 qr. in pounds and ounces.

52. Reduce 3.327 qrs. to pounds.

53. Reduce 4.684£. to pence.

54. Find the value of .346 of a day in hours, minutes, &c.

55. Find the value of .5876 of an hour in minutes and seconds.

56. Express in decimals of a foot the inches in the following numbers :—3 ft. 6 in.; 4 ft. 3 in.; 7 ft. 9 in.; 3 ft. 8 in.; 5 ft. 7 in.; 9 ft. 10 in.

57. Find the value of .375 ft. in inches and parts.

58. Find the value of .468 of a square foot in square inches.

59. Find the value of .8438 of a solid foot in solid inches.

60. How many square feet in a board 9 in. wide and 15 ft. 3 in. long.

Change the inches to decimals of a foot. Since the answer will be in square feet, it will be necessary to find the value of the decimal in square inches. In general, however, it will be quite as convenient to let the answer remain in decimals. The answer is 11.4375 ft. It will be sufficiently exact to call it 11.4 ft.

61. How many square feet in a floor 14 ft. 7 in. wide and 19 ft. 4 in. long ?

62. How many square feet in a board 1 ft. 8 in. wide and 17 ft. 10 in. long.

63. How many solid feet in a stick of timber 28 ft. 4 in. long. 1 ft. 2 in. wide, and 11 in. deep ?

Note. In questions of this kind it will generally be most convenient to change the inches to, decimals of a foot, because when the who . for 1 year, what would be t...bers per come very large and for 3 years ? For 4 years ? generally, and hundredths in alm what would be the rate per ficiently exact for common purposes. For 4 months ? For timber, boards, wood, &c. would find it ext ? For 7 months ? to have their rules divided into tenths of a foot, instead of inches.

There is a method of performing examples of this kind called *duodecimals*, which will be explained hereafter, but it is not so convenient as decimals. .

64. How many solid feet in a pile of wood 4 ft. 2 in. wide, 3 ft. 8 in. high, and 13 ft. 4 in. long ?

It has been already remarked that in interest, discount

commissions, &c. 6 per cent., 7 per cent., &c. signifies $\frac{6}{100}$, $\frac{7}{100}$, &c. of the sum. This may be written as a decimal fraction. In fact this is the most proper and the most convenient way to express, and to use it. 1 per cent. is .01; 2 per cent. is .02; 6 per cent. is .06; 15 per cent. is .15; 6½ per cent. is .065, &c. This manner of expressing the rate will be very simple in practice, if care be taken to point the decimals right in the result.

65. A commission merchant sold a quantity of goods amounting to $583.47, for which he was to receive a commission of 4 per cent. How much was the amount of the commission ?

$$583.47$$
$$.04$$
$$\overline{}$$

$23.3388 *Ans.*

There are two decimal places in each factor, consequently there must be four places in the result. The answer is $23.34 nearly.

66. What is the commission on $1358.27, at 7 per cent. ?
67. What is the commission on $1783.425, at 5 per cent. ?
68. A merchant bought a quantity of goods for $387.48, and sold them so as to gain 15 per cent. How much did he gain, and for how much did he sell the goods ?
69. What is the insurance of a ship and cargo, worth $53250, at 2½ per cent. ?

Note. 2½ per cent is equal to .025, for 2 per cent. is .02, and ½ per cent. is ½ of an hundredth, which is 5 thousandths.

70. What books, of which the
43. How many cwt. of cotton in 5⅔ ba...
taining 4 cwt. 3 qrs. 7 lb. ?
44. How many cwt of custom-house to add $\frac{1}{10}$ or 10
taining 1 cwt. 3 qrs. before casting the duties. 10 per
Find the is $15.737, which, added to $157.37
makes $173.107. The duties must be reckoned on $173.107.
When the duties are stated at 15 per cent. they will actually
be 16½ per cent. on the invoice; because 15 per cent. on
$\frac{1}{10}$ will amount to 1½ per cent. on the whole. It will be
most convenient generally to reckon the duties at 16½ per
cent., instead of adding $\frac{1}{10}$ of the sum and then reckoning
them at 15 per cent. When the duties are at any other rate,
the rate may be increased $\frac{1}{10}$ of itself, instead of increasing

e invoice $\frac{1}{10}$. For instance, if the rate is 10 per cent. call
11 per cent., if the rate is 14 per cent. call it $15\frac{4}{10}$ per
cent., then the multiplier will be .154. If the rate is $12\frac{1}{2}$
per cent., that is, .125, $\frac{1}{10}$ of this is .0125, which added to
.125 makes .1375 for the multiplier.

71. What is the duty on a quantity of tea, of which the
invoice is $215.17, at 50 per cent. ?

72. What is the duty on a quantity of wine, of which the
invoice is $872 at 40 per cent. ?

73. What is the duty on a quantity of saltpetre, of which
the invoice is $1157, at 7½ per cent. ?

74. Imported a quantity of hemp, the invoice of which
was $1850, the duties 13¼ per cent. What did the hemp
amount to after the duties were paid ?

75. Bought a quantity of goods for $58.43, but for cash
the seller made a discount of 20 per cent. What did the
goods amount to after the discount was made ?

76. A merchant bought a quantity of sugar for $183.58,
but being damaged he sold it so as to lose 7½ per cent.
How much did he sell it for ?

77. Bought a book for $.75, but for cash a discount of
20 per cent. was made. What did the book cost ?

78. Bought a book for $4.375, but for cash a discount
of 15 per cent. was made. How much did the book cost ?

79. What is the interest of $43.25 for 1 year, at 6 per
cent. ?

80. What is the interest of $183.58 for 1 year at 7 per
cent. ?

81. At 6 per cent. for 1 year, what would be the rate per
cent. for 2 years ? For 3 years ? For 4 years ?

82. At 6 per cent. for 1 year, what would be the rate per
cent. for 6 months ? For 2 months ? For 4 months ? For
1 month ? For 3 months ? For 5 months ? For 7 months ?
For 8 months ? For 9 months ? For 10 months ? For 11
months ?

83. At 6 per cent. for 1 year, what would be the rate per
cent. for 13 months ? For 14 months ? For 1 year and 5
months ?

84. If the rate for 60 days is 1 per cent., or .01, what is
the rate for 6 days ? For 12 days ? For 18 days ? For
24 days ? For 36 days ? For 42 days ? For 48 days ? For
54 days ?

Note. The interest of 6 days is $\frac{1}{10}$ per cent., that is .001. The interest of 1 day therefore will be $\frac{1}{6}$ of $\frac{1}{10}$, or $\frac{1}{60}$ per cent., or .00016. The rate for 2 days twice as much, &c. In fact the rate for the days may always be found by dividing the number of days by 6, annexing zeros if necessary, and placing the first figure in the place of thousandths, if the number of days exceeds 6.

85. What is the interest of $47.23 for 2 months, at 6 per cent. ?

Note. When the rate per cent is stated without mentioning the time, it is to be understood of 1 year, as in the following examples.

86. What is the interest of $27.19 for 4 months, at 6 per cent. ?

87. What is the interest of $147.96 for 6 months, at 6 per cent. ?

88. What is the interest of $87.875 for 8 months, at 6 per cent. ?

89. What is the interest of $243.23 for 14 months, at 6 per cent. ?

90. What is the interest of $284.85 for 3 months, at 6 per cent. ?

91. What is the interest of $28.14 for 5 months, at 6 per cent. ?

92. What is the interest of $12.18 for 7 months, at 6 per cent. ?

93. What is the interest of $4.38 for 9 months, at 6 per cent. ?

94. What is the interest of $15.125 for 11 months, at 6 per cent. ?

95. What is the interest of $127.47 for 2 months and 12 days, at 6 per cent. ?

96. What is the interest of $873.62 for 4 months and 24 days, at 6 per cent. ?

97. What is the interest of $115.42 for 7 months and 15 days, at 6 per cent. ?

98. What is the interest of $516.20 for 11 months and 23 days, at 6 per cent. ?

99. What is the interest of $143.18 for 1 year, 7 months, and 14 days, at 6 per cent. ?

100. A gave B a note for $357.68 on the 13th Nov. 1819, and paid it on the 11th April, 1822, interest at 6

per cent. How much was the principal and interest together at the time of payment ?

101. A note for $843.43 was given 5th July, 1817, and paid 14th April, 1822, interest at 6 per cent. How much did the principal and interest amount to ?

102. A note was given 7th March, 1818, for $587; a payment was made 19th May, 1819, of $53, and the rest was paid 11th Jan. 1820. What was the interest on the note ?

103. What is the interest of $157 for 2 years, at 5 per cent. ?

104. What is the interest of 13£. 3s. 6d. for 1 year, at 6 per cent. ?   ·) .

Note. If the shillings be reduced to a decimal of a pound, the operation will be as simple as on Federal money. The following is a more simple method of changing shillings to decimals, than the one given above. $\frac{1}{10}$ part of 20s. is 2s., therefore every 2s. is $\frac{1}{10}$£. or .1£. Every shilling is $\frac{1}{20}$£., that is $\frac{5}{100}$£. or .05£. ; 3s. then is .1£. and .05£., or .15£.

In 1£. there are 960 farthings. 1 farthing then is $\frac{1}{960}$ of 1£. 6d. is 24 farthings, consequently $\frac{24}{960}$ of a £. These are rather larger than thousandths, but they are so near thousandths that in small numbers they may be used as thousandths. $\frac{24}{960}$£.$=\frac{1}{40}$£. when reduced, and $\frac{25}{1000}$£.$=\frac{1}{40}$£., so that 24 farthings are exactly $\frac{25}{1000}$£. or .025£. If the number of farthings is 13 they will be $\frac{13}{1000}$£. and rather more than $\frac{1}{4}$ of another thousandth. This may be called $\frac{14}{1000}$ or .014, and the error will be less than $\frac{1}{4}$ of $\frac{1}{1000}$. If the number of farthings be less than 12 they may be called so many thousandths, and the error will be less than $\frac{1}{4}$ of $\frac{1}{1000}$. If the number of farthings is between 12 and 36 add 1 to them, if more than 36 add 2, and call them so many thousandths; and the result will be correct within less than $\frac{1}{4}$ of $\frac{1}{1000}$. 48 farthings make 1 shilling, therefore there will never be occasion to use more than this number. From the above observations we obtain the following rule. *Call every two shillings one tenth of a pound, every odd shilling five hundredths, and the number of farthings in the pence and farthings so many thousandths, adding one if the number is between twelve and thirty-six, and two if more than thirty-six.*

It will be well to remember this rule, because it will be

useful in many instances, particularly in changing English money to dollars and cents, and the contrary.

13£. 3s. 6d. then is reduced as follows : 2s.=.1£. 1s.= .05£. and 6d.=24 farthings=.025£. and the whole is equal to £13.175.

$$13.175$$
$$.06$$
_____

£ .79050 *Ans.*

To change the result to shillings and pence it is necessary to reverse the above operation. The .7 or $\frac{7}{10}$ are 14s. The .09 or $\frac{9}{100}$ are $\frac{5}{100}+\frac{4}{100}$. The $\frac{5}{100}$ are 1s. and the $\frac{4}{100}$ are $\frac{40}{1000}$, or 40 farthings ; then taking out 2, because the number is above 36, we have 38 farthings, or 9d. 2qr. ; and the whole interest is 15s. 9d. 2qr.

105. What is the interest of 13£. 15s. 3d. 2 qr. for 1 year and 6 months, at 6 per cent. ?

106. What is the interest of 4£. 11s. 8d. 1qr. for 9 months and 15 days, at 6 per cent. ?

107. What is the interest of 137£. 0s. 9d. from 13th May, 1811, to 19th July, 1815, at 6 per cent. ?

108. What is the interest of 137£. 17s. 2d. from 11th Jan. 1822, to 15th August, at 6 per cent. ?

109. What is the interest of 17£. 9s. from 1st June, 1819, to 17th Aug. 1820, at 6 per cent. ?

110. What is the interest of 13s. 4d. from 17th June, 1818, to 28th Aug. 1821, at 6 per cent. ?

111. What is the interest of 4s. 8d. 2qr. for 7 months and 3 days, at 6 per cent. ?

112. What is the commission on 143£. 13s., at 5 per cent. ?

113. What is the duty on a quantity of goods, of which the invoice is 257£. 19s. 4d., at 15 per cent. ?

N. B. The above examples in pounds, shillings, &c. apply equally to English and to American money.

Division of Decimals.

XXVIII. 1. If 5 barrels of cider cost $18.75, what is that per barrel ?

2. A man bought 17 sheep for $98.29, what was the average price ?

3. Divide $183.575 equally among 5 men. How much will each have ?

4. Divide 7.5 barrels of flour equally among 5 men. How much will they have apiece ?

5. Divide 11.25 bushels of corn equally among 8 men. How much will they have apiece ?

6. A man travelled 73.487 miles in 15 hours; what was the average distance per hour ?

7. At 28£. 5s. 8d. per ton, what cost 1 cwt. of iron ?

8. If a ship and cargo are worth 1253£. 6s. 4d., what is the man's share who owns $\frac{1}{15}$ of her ?

9. What is $\frac{1}{8}$ of 49.376 ?

10. What is $\frac{1}{17}$ of 583.542 ?

11. What is $\frac{1}{39}$ of 13.75 ?

12. What is $\frac{1}{135}$ of 387.65 ?

13. Divide 13.8468 by 4.

14. Divide 1387.35 by 48.

15. Divide 158.6364 by 113.

16. Divide 12.4683 by 27.

17. Divide 1.384 by 15.

18. Divide .7376 by 28.

19. Divide .6438 by 156.

20. Divide 1.5 by 58.

21. Divide .4 by 13.

22. Divide .0346 by 27.

23. Divide .003 by 43.

24. Divide 1.06438 by 1846.

25. Divide 13.84783 by 137648.

26. At $1.37 per gallon, how many gallons of wine may be bought for $37 ?

27. At $.34 per bushel, how many bushels of oats may be bought for $24 ?

29. At $.165 per lb., how many lb. of raisins may be bought for $3 ?

30. At $.03 apiece, how many lemons may be bought for $5 ?

31. If 1.75 yards of cloth will make a coat, how many coats may be made from 38 yards ?

32. If 1.3 bushels of rye is sufficient to sow an acre of ground, how many acres will 23 bushels sow ?

33. If 18.75 bushels of wheat grow on 1 acre, how many acres will produce 198 bushels, at that rate ?

34. If a man travel 5.385 miles in an hour, in how many hours will he travel 83 miles at that rate ?

9

35. If 3s. will pay for 1 day's work, how many days' work may be had for 13s. ?

36. If 5s. 8d. will pay for 1 day's work, how many days' work will 11£. pay for ?

37. At 8s. 3d. per gallon, how many gallons of wine may be bought for 18£. ?

38. If 2.5 barrels of cider cost $7, what is that per barrel ?

39. If 1.5 barrel of flour cost $10, what is that per barrel ?

40. If 2.75 firkins of butter cost $23, what is that per firkin ?

41. If 3.375 barrels of beer cost $14, what is that per barrel ?

42. If 13.16 bushels of wheat cost 6£., what is that per bushel ?

43. If .3 of a yard of cloth cost $2, what is that per yard ?

44. If .35 of a ton of hay cost $8, what cost a ton ?

45. If .846 of a barrel of flour cost 32 shillings, what will a barrel cost at that rate ?

46. If .137 of a ton of iron cost 52 shillings, what will 1 ton cost ?

47. How many times is 1.3 contained in 18 ?

48. How many times is 3.25 contained in 39 ?

49. How many times is 4.75 contained in 180 ?

50. How many times is 16.375 contained in 4,876 ?

51. How many times is 24.538 contained in 63 ?

52. How many times is 1.372 contained in · 14 ?

53. How many times is 4.1357 contained in 15 ?

54. How many times is .3 contained in 3 ?

55. How many times is .04 contained in 4 ?

56. How many times is .13 contained in 8 ?

57. How many times is .385 contained in 17 ?

58. How many times is .0684 contained in 47 ?

59. How many times is .0001 contained in 53 ?

60. How many times is .0005 contained in 127 ?

61. 3 is .3 of what number ?

62. 4 is .04 of what number ?

63. 8 is .13 of what number ?

64. 17 is .385 of what number ?

65. 47 is .0684 of what number ?

66. 53 is .0001 of what number ?

67. 127 is .0005 of what number?

68. How many times is .0035 contained in 67?
69. 67 is .0035 of what number?
70. Divide 156 by 4.35.
71. Divide 38 by 13.56.
72. Divide 23 by 1.3846.
73. Divide 7 by 8.4.
74. 7 is what part of 8.4?
75. Divide 3 by 5.8.
76. 3 is what part of 5.8?
77. Divide 8 by 17.37.
78. 8 is what part of 17.37?
79. Divide 23 by 120.684.
80. 23 is what part of 120.684?
81. Divide 14 by .7.
82. Divide 130 by .83.
83. Divide 847 by .134.
84. Divide 8 by .0645.
85. Divide 3 by .00735.
86. Divide 1 by .005643.
87. Divide 157 by .00001.
88. At $2.75 per gallon, how many gallons of wine may be bought for $56.03?
89. At 17.375 shillings per gallon, how many gallons of wine may be bought for 42.25 shillings?
90. At 16s. 4d. per gallon, how many gallons of brandy may be bought for 4£. 7s.?
91. At 2£. 3s. 4d. per barrel, how many barrels of flour may be bought for 32£. 7s. 6d.?
92. If 3.75 barrels of flour cost $25.37, how much is that per barrel?
93. If 5.375 barrels of cider cost 4£. 4s., what is that per barrel?
94. If .845 of a yard of cloth cost $5.37, what is that per yard?
95. If $\frac{4}{5}$ of a ton of iron cost $60.45, what cost 1 ton?
96. How many times is 13.753 contained in 42.7?
97. How many times is 1.468 contained in 473.75?
98. How many times is .7647 contained in 13.42?
99. How many times is .0738 contained in 1.6473?
100. 1.6473 is .0738 of what number?
101. How many times is .001 contained in .1?
102. .1 is .001 of what number?
103. How many times is .002 contained in .01?

104. .01 is .002 of what number?
105. How many times is .002 contained in .002?
106. .002 is .002 of what number?
107. Divide 31.643 by 2.3846.
108. Divide 2.4637 by .6847.
109. If 1 lb. of candles cost $.14, how many lb. may be bought for $1.375?
110. If 4.5 yards of cloth cost $28.35, how much is that per yard?
111. If 3.45 tons of hay cost 22£. 7s. 5d., how much is that per ton?
112. At 3s. 8d. per bushel, how many bushels of barley may be bought for 3£. 5s. 7d.?
113. If 47.25 bushels of barley cost 15£. 17s. 5d., what is that per bushel?
114. If 15 cwt. 3 qr. 14 lb. of iron cost 17£. 14s. 8d., what is that per cwt.?
115. If .35 of a ton of iron cost 10£. 3s. 5d., what cost a ton at that rate?
116. Divide 16.4567 by 2.5.
116. Divide 137.06435 by 3.25.
117. Divide 105.738 by .3.
118. Divide 75.426 by .1.
119. Divide 1.76453 by 1.3758.
120. Divide .78357 by .001.
121. Divide .073467 by .005.
122. Divide .007468 by .0075.
123. How many times is .037 contained in 1.04738?
124. 1.04738 is .037 of what number?
125. How many times is .135 contained in 13.4073?
126. 13.4073 is .135 of what number?
127. Divide 13.40764 by 123.725.
128. Divide .406478 by 135.407.

In the following examples express the division in the form of a common fraction, and reduce them to their lowest terms.

129. Divide 17.57 by 14.23.
130. Divide 3.756 by 5.873.
131. Divide .6375 by .5268.
132. Divide 3.45 by 2.756.
133. Divide 1.6487 by 2.35.
134. Divide 113.45 by 21.4764.
135. Divide .7384 by .37.

136 Divide .007 by .5.
137. Divide .647387 by .0042.
138. Divide .53 by .00067.
139. Divide .003 by 0.00001.
140. 3.5 is what part of 7.8 ?
141. 13.76 is what part of 17.5 ?
142. 7.0387 is what part of 42.95 ?
143. 1.5064 is what part of 8.944783 ?

Miscellaneous Examples.

1. If 1.4 cwt. of sugar cost $10.09, what will 9 cwt. 3 qrs. cost ?

2. If 19¼ yards of cloth cost $128.35, what will 18 yds. 3 qrs. cost ?

3. If 23½ yds. of riband cost $5¼, what will 34½ yds. cost ?

4. If 3 cwt. 2 qrs. 14 lb. of sugar cost $38.55, what will 19 cwt. 1 qr. 17 lb. cost ?

5. If ½ cwt. of tobacco cost 4£. 18s., how much may be bought for 13£. 17s. 8d. ?

6. Sold 75¾ chaldrons of lime, at 11s. 6d. per chaldron, how much did it come to ?

7. A goldsmith sold a tankard for 10£. 13s., at the rate of 5s. 6d. per oz.; how much did it weigh ?

8. Goliah the Philistine is said to have been 6½ cubits high, each cubit being 1 ft. 7.168 English inches ; what was his height in English feet ?

9. How many yards of flannel that is 1 English ell wide will be sufficient to line a cloak containing 8½ yds., that is ¾ yd. wide ?

10. I agreed for a carriage of 2.5 tons of goods 2.9 miles, for .75 of a guinea ; what is that per cwt. for 1 mile ?

11. If a traveller perform a journey in 35.3 days, when the days are 11.374 hours long ; in how many days will he perform it, when the days are 9.13 hours long ?

12. If 12 men can do 125 rods of ditching in 65¾ days; in how many days can they do $242\frac{4}{13}$ rods ?

13. If a room 18 ft. 6 in. long, and 14 ft. 9 in. wide, how many square feet ? In a yard of carpeting that is 2 ft. 8 in. wide, how many square feet ? How many yards of such carpeting will cover the above mentioned floor ?

9 *

14. How many yards of carpeting that is 1¼ yd. wide will cover a floor 22 ft. 7 in. long, and 19 ft. 8 in. wide ?

15. How many feet of boards will it take to cover the walls of a house 32 ft. 8 in. long, 26 ft. 4 in. wide, and 26 ft. 5 in. high ?   How much will they cost at $3.50 per 1000 feet ?

16. How many feet will it take to cover the floors of the above house ?

17. If 1000, or a bunch, of shingles will cover 10 feet square, how many bunches will it take to cover the roof of the above house, allowing the length of the rafters to be 16 ft. 5 in. ?

18. In a piece of land 37¾ rods long, and 32⅘ rods wide, how many acres ?

19. What will a piece of land, measuring 57 ft. in length, and 43 ft. in breadth, come to, at the rate of $25 per square rod ?

20. In a pile of wood 23 ft. 7 in. long, 3 ft. 10 in. wide, and 4 ft. 3 in. high, how many cords ?

21. How many feet of wood in a load 8 ft. long, 4 ft. wide, and 3 ft. 8 in. high ?

N. B.   Wood prepared for the market is generally 4 feet long, and a load in a wagon generally contains two lengths, or 8 feet in length.   If a load is 4 feet high and 4 feet wide, it contains a cord.   It was remarked above, that what is called one foot of wood, is 16 solid feet, and that 8 such feet make 1 cord.   To find how many of these feet a pile or load of wood contains, it is necessary to find the number of solid feet in it, and then to divide by 16.   When the load of wood is 8 feet long, we may multiply the breadth and height together, and then, instead of multiplying by 8, and dividing by 16, we may divide at first by 2, and the same result will be obtained.

22. How many feet of wood in a load 8 feet long, 3 ft. 4 in. wide, and 2 ft. 7 in. high ?

23. How many feet of wood in a load 8 feet long. 3 ft. 7 in. wide, and 5 ft. 2 in. high ?

24. How much wood in a load 8 ft. long, 4 ft. 2 in. wide, and 5 ft. 4 in. high ?

25. If a load of wood is 8 ft. long, and 3 ft. 7 in. wide, how high must it be to make a cord ?

26. How many bricks 8 inches long, 4 inches wide, and

$2\frac{1}{2}$ inches thick, will it take to build a house 44 feet long, 40 feet wide, 20 feet high, and the walls 12 inches thick?

27. What is the value of 87 pigs of lead, each weighing 3 cwt. 2 qrs. $17\frac{1}{4}$ lb., at 8£. 13s. 8d. per fother of $19\frac{1}{4}$ cwt.?

28. What is the tax upon $1153. at $.03 on a dollar?

29. What is the tax upon $843.35, at $.04 on a dollar?

30. What is the tax upon 785£. 11s. 4d. at 2s. 5d. on the pound?

· 31. Suppose a certain town is to pay a tax of $6145.88, and the whole property of the town is valued at $153647; what is that on a dollar? How much must a man pay, whose property is valued at $23475.67?

Note. In assessing taxes, the first requisite is to have an inventory of the property, both real and personal, of the whole town or parish, and also of each individual who is to be taxed, and the number of polls. The polls are always stated at a certain rate. Then knowing the whole tax, take out what the polls amount to, and the remainder is to be laid upon the property. Find how much each dollar is to pay, and make a table, containing the portion for 1, 2, 3, &c. to 10 dollars, *hen for 20, 30, 40, &c. to 100, and then for 200, 300, &c. From this table it will be easy to find the tax upon the property of any individual.

32. A certain town is taxed $3137.43. The whole property of the town is valued at $89640.76. There are 120 polls which are taxed $.75 each. What is the tax on a dollar? How much is a man's tax who pays for 3 polls, and whose property is valued at $2507?

33. A merchant bought wine for $1.75 per gallon, and sold it for $2.25 per gallon. What per cent. did he gain?

Note. He gained 50 cents on a gallon, which is $\frac{50}{175}=\frac{10}{35}$ of the first cost. It has been already remarked that 1 per cent. is .01, 2 per cent. is .02, &c.; that is, the rate per cent. is always a decimal fraction carried to two places or hundredths. To find the rate per cent. then, first make a common fraction, and then change it to a decimal $\frac{10}{35}=.285$. Now .28 is 28 per cent. and .0055 is$\frac{55}{100}$per cent. The rate then $28\frac{55}{100}$ per cent. The two first decimal places taken together being hundredths are so much per cent., and thousandths are so many tenths of one per cent.

34. A merchant bought a hhd. of molasses for $20, and sold it for $25; what per cent. did he gain?

35. A merchant bought a quantity of flour for $137, and sold it for $143 ; what per cent. did he gain ?

36. A man bought a quantity of goods for $94.37, and sold them for $83.92. What did he lose per cent. ?

37. A merchant bought molasses for 1s. 8d. per gallon, and sold it for 2s. 3d. per gallon. What did he gain per cent ?

38. A merchant bought wine for 11s. 3d. per gallon, and sold it for 9s. 8½d. What per cent. did he lose ?

39. A merchant bought a quantity of goods for 37£. 15s 8d. ; and sold them again for 43£. 11s. 4d. What per cent did he gain ?

40. A man buys a quantity of goods for $843 ; what per cent. profit must he make in order to gain $157 ?

41. A man failing in trade owes $19137.43, and his property is valued at $13472.19. What per cent. can he pay ?

42. A man purchased a quantity of goods, the price of which was $57, but a discount being made, he paid $45.60. What per cent. was the discount ?

43. A man hired $87 for 1 year, and then paid for principal and interest $92.22. What was the rate of the interest ?

44. A man paid $12.81 interest for $183, for 2 years. What was the rate per year ?

45. A man paid $13.125 interest for $135, for 1 year and 6 months. What was the rate per year ?

46. A man paid $4.37 interest for $58, for 1 year and 8 months. What was the rate per year ?

47. 4s. 6d. sterling of England is equal to 1 dollar in the United States. What is the value of 1£. sterling in Federal money ?

48. How many dollars in 35£. sterling ?

49. How many dollars in 27£. 14s. 8d. ?

Note. Change the shillings and pence to the decimal of a pound, by the short method shown above.

50. How many dollars in 187£. 17s. 4d. ?

51. In $19.42 how many pounds sterling ?

52. In $157 how many pounds ?

53. In $2384.72 how many pounds ?

54. Bought goods in England to the amount of 123£. 17s. 9d. ; expenses for getting on board 3£. 5s. 8d. ; $8.50 freight ; duties in Boston 15 per cent. on the invoice ; other expenses in Boston $15.75. How many dollars did the goods cost ? How much must they be sold for to gain 12 *per cent.* on the cost ?

55. What is the interest of $47,50 for 1 year, 7 months, and 13 days, at 7 per cent. ?

```
        47.50
          .07
       ------
       3.3250  Interest for 1 year.
       1.6625     do.    for 6 months.
        .277+     do.    for 1 month.
        .092+     do.    for 10 days.
        .03 nearly do.   for 3 days.
       ------
  Ans. 5.3865
```

I first find the interest for 1 year, and then $\frac{1}{2}$ of that is the interest for 6 months; $\frac{1}{6}$ of the interest for 6 months will be the interest for 1 month; $\frac{1}{3}$ of the interest for 1 month will be the interest for 10 days, and $\frac{1}{3}$ of the interest for 10 days is very near the interest for 3 days. All these added together will give the interest for the whole time. In a similar manner, the interest for any time at any rate per cent. may be calculated.

When there are months and days, it is better to calculate the interest first at 6 or 12 per cent., and then change it to the rate required. Observe that 1 per cent. is $\frac{1}{6}$ of 6 per cent., $1\frac{1}{2}$ per cent. is $\frac{1}{4}$ of 6 per cent., 2 per cent is $\frac{1}{3}$ of 6 per cent, &c. Hence if the rate is 7 per cent., calculate first at 6 per cent., and then add $\frac{1}{6}$ of it to itself, or if 5 per cent., subtract $\frac{1}{6}$; if $7\frac{1}{2}$ or $4\frac{1}{2}$ per cent. add or subtract $\frac{1}{4}$, &c.

Let us take the above example.

6 per cent. for 1 year, 7 months, and 13 days, is $9\frac{7}{10}$ per cent. nearly, that is .097.

```
            47.50
             .097
          -------
            33250
           42750
          -------
  1/6 of 4.60750  Interest at 6 per cent.
            7679     do.   at 1 per cent.
          -------
         $5.3754
```

This answer agrees with the other within about 1 cent. Greater accuracy might be attained, by carrying the rate to one or two more decimal places.

56. What is the interest of $135.16 from the 4th June, 1817 to 13th April, 1818, at 5 per cent. ?

57. What is the interest of $85.37 from 18th July, 1815, to 17th Nov. 1818, at 4½ per cent. ?

58. What is the interest of $45.87 from 19th Sept. 1819, to 11th Aug. 1821, at 7½ per cent. ?

59. What is the interest of $183 from 23d Oct. 1817, to 19th Jan. 1820, at 4 per cent. ?

60. What is the interest of 113£. 14s. for 1 year, 5 months, and 8 days, at 7 per cent. ?

61. What is the interest of 87£. 15s. 4d. for 2 years, 11 months, 3 days, at 7½ per cent. ?

62. What is the interest of 43£. 16s. for 9 months and 13 days, at 8 per cent. ?

63. What is the interest of 142£. 19s. for 1 year, 8 months, and 13 days, at 9 per cent. ?

64. What is the interest of $372 for 4 years, 8 months, and 17 days, at 7¼ per cent. ?

65. What is the interest of 1 dollar for 15 days at 7 per cent. ?

66. What is the interest of $.25 for 13 days, at 7½ per cent. ?

67. What is the interest of $.375 for 19 days, at 11 per cent. ?

68. What is the interest of $1147 for 8 hours, at 6 per cent. ?

69. What is the interest of 137£. 11s. for 11 days at 9 per cent. ?

70. What is the interest of 15s. for 3 months, at 8 per cent. ?

71. What is the interest of 16£. 7s. 8d. for 2 months, at 12 per cent. ?

72. What is the interest of 4s. 3d. for 17 years, 3 months, and 7 days, at 8 per cent. ?

73. A man gave a note 13th Feb. 1817, for $753, interest at 6 per cent., and paid on it as follows: 19th. Aug. 1817, $45 ; 27th June, 1818, $143 ; 19th Dec. 1818, $25 ; 11th May 1819, $100 ; and 14th Sept. 1820, he paid the rest, principal and interest. How much was the last payment ?

74. A note was given 17th July, 1814, for $1432, interest at 6 per cent., and payments were made as follows ; 15th Sept. same year, $150 ; 2d Jan. 1815, $130 ; 16th. Nov. 1815, $23 ; 11th April, 1817, $237 ; 15th Aug. 1818, $47. *How much* was due on the note, principal and interest, 5th Feb. 1819 ?

ARITHMETIC.

PART II.

NUMERATION.

I. A single thing of any kind is called a *unit* or *unity*.

Particular names are given to the different collections of units.

A single unit is called - - - - - - *One.*

If to one unit we join one unit more, the collection is called *two*; that is, *one* added to *one* is called *two*, or one and one are - - - - - - - - - - *Two.*

One added to *two* is called *three*; two and one are *Three.*

One added *three* is called *four*; three and one are *Four.*

One added to *four* is called *five*; four and one are *Five.*

One added to *five* is called *six*; five and one are *Six.*

One added to *six* is called *seven*; six and one are *Seven.*

One added to *seven* is called *eight*; seven and one are - - - - - - - - - - - - *Eight.*

One added to *eight* is called *nine*; eight and one are *Nine.*

One added to *nine* is called *ten*; nine and one are *Ten.*

In this manner we might continue to add units, and to give a name to each different collection. But it is easy to perceive that, if it were continued to a great extent, it would be absolutely impossible to remember the different names; and it would also be impossible to perform operations on large numbers. Besides, we must necessarily stop somewhere; and at whatever number we stop, it would still be possible to add more; and should we ever have occasion to do so, we should be obliged to invent new names for them, and to explain them to others. To avoid these inconveniences, a method has been contrived to express all the numbers, that are necessary to be used, with very few names.

The first ten numbers have each a distinct name. The collection of ten simple units is then considered a unit : it is called a unit of the *second order.* We speak of the collections of ten, in the same manner that we speak of simple units ; thus we say one ten, two tens, three tens, four tens, five tens, six tens, seven tens, eight tens, nine tens. These expressions are usually contracted ; and instead of them we say ten, twenty, thirty, forty, fifty, sixty, seventy, eighty, ninety.

The numbers between the tens are expressed by adding the numbers below ten to the tens. One added to ten is called ten and one ; two added to ten is called ten and two; three added to ten is called ten and three, &c. These are contracted in common language ; instead of saying ten and three, ten and four, &c., we say thirteen, fourteen, fifteen, sixteen, seventeen, eighteen, nineteen. These names seem to have been formed from three and ten, four and ten, &c. rather than from ten and three, ten and four, &c., the number which is added to ten being expressed first. The signification, however, is the same. The names eleven and twelve, seem not to have been derived from *one and ten, two and ten ;* although twelve seems to bear some analogy to two. The names *oneteen, twoteen,* would have been more expressive ; and perhaps all the numbers from ten to twenty would be better expressed by saying *ten one, ten two, ten three,* &c.

The numbers between twenty and thirty, and between thirty and forty, &c. are expressed by adding the numbers below ten to these numbers; thus one added to twenty is called twenty-one, two added to twenty is called twenty-two, &c. ; one added to thirty is called thirty-one, two added to thirty is called thirty-two, &c.; and in the same manner forty-one, forty-two, fifty-one, fifty-two, &c. All the numbers are expressed in this way as far as ninety-nine, that is nine tens and nine units.

If one be added to ninety-nine, we have ten tens. We then put the ten tens together as we did the ten units, and this collection we call a unit of the *third order*, and give it a name. It is called one *hundred.*

We say one hundred, two hundreds, &c. to nine hundreds, in the same manner, as we say one, two, three, &c.

The numbers between the hundreds are expressed by adding tens and units. With units, tens, and hundreds we can express nine hundreds, nine tens, and nine units ; which is called nine hundred and ninety-nine.   If one unit be added to this number, we have a collection of ten hundreds ; this is also made a unit, which is called a unit of the *fourth order ;* and has a name.   The name is *thousand.*

This principle may be continued to any extent.   Every collection of ten units of one order is made a unit of a higher order ; and the intermediate numbers are expressed by the units of the inferior orders.   Hence it appears that a very few names serve to express all the different numbers which we ever have occasion to use.   To express all the numbers from one to nine thousand, nine hundred, and ninety-nine, requires, properly speaking, but twelve different names.   It will be shown hereafter, that these twelve names express the numbers a great deal farther.

Various methods have been invented for writing numbers, which are more expeditious, than that of writing their names at length, and which, at the same time, facilitate the pro-cesses of calculation.   Of these the most remarkable is the one in common use, in which the numbers are expressed by characters called *figures.*   This method is so perfect, that no better can be expected or even desired.   These figures are supposed to have been invented by the Arabs ; hence they are sometimes called Arabic figures.   The figures are nine in number.   They are exactly accommodated to the manner of naming numbers explained above.[*]

[*] Next to the Arabic figures, the Roman method seems to be the most convenient and the most simple.   It is very nearly accommodat-ed to the mode of naming numbers explained above.   A short descrip-tion of it may be interesting to some ; and it will often be found ex-tremely useful to explain this method to the pupil before the other. The pupil will understand the principles of this, sooner than of the other, and having learned this, he will more easily comprehend the other.   He will perfectly comprehend the principle of carrying, in this, both in addition and subtraction, and the similarity of this to the com-mon method is so striking that he will readily understand that also.

The pupil may perform some of the examples in Sects. I, II, and VIII, Part I, with Roman characters.

THE ROMAN NOTATION.

One was written with a single mark, thus,	I
Two was written with two marks . .	II
Three was written	III
Four was written	IIII

One is written   -   -   1
Two is written   2
Three is written   -   -   -   3
Four is written   -   -   4
Five is written   -   -   -   5
Six is written   6
Seven is written   -   -   -   7
Eight is written   -   -   8
Nine is written   -   -   -   9

These nine figures are sometimes called the 9 digits.   **By**

Five was written   IIIII
Six was written   IIIIII
Seven was written   IIIIIII
Eight was written   IIIIIIII
Nine was written   IIIIIIIII
Ten, instead of being written with ten marks,
    was expressed by two marks crossing each
    other, thus, ... X
    which expressed a unit of the second order.
Two tens or twenty were written   .   . X X
Three tens or thirty were written   .   . X X X

And so on to ten tens, which were written with ten crosses. But as it was found inconvenient to express numbers so large as seven or eight, with marks as represented above, the X was cut in two, thus X, and the upper part V was used to express one half of ten, or five, and the numbers from five to ten were expressed by writing marks after the V, to express the number of units added to five.

Six was written   .   .   .   .   . V I
Seven was written   .   .   .   . V I I
Eight was written   .   .   . V I I I
Nine was written   .   .   .   .   . V I I I I

The intermediate numbers between the tens were expressed by writing the excess above even tens after the tens.

Eleven was written   .   .   .   . X I
Twelve was written   .   .   .   . X I I, &c.
Twenty-seven was written   .   . X X V I I, &c.

To express ten Xs, or ten tens, that is, one unit, of the third order, or one hundred, three marks were used, thus, C. And to avoid the inconvenience of writing seven or eight Xs, the C was divided, thus C, and the lower part L used to express five Xs, or fifty.

To express ten hundreds, four dashes were used, thus, M. This last was afterwards written in this form CƆ and sometimes CIƆ, and was then divided, and IƆ was used to express five hundreds.

These dashes resemble some of the letters of the alphabet, and those letters were afterwards substituted for them.

The I resembles the I; the V resembles the V; the X resembles the X, the L resembles the L; the C was substituted for the C; the IƆ resembles the D; and the M resembles the M.

these nine characters all numbers whatever may be expressed.

To express ten, we make use of the first character 1. But
to distinguish it from one unit, it is written in a new place,
thus 10 ; the 0, which is called *zero* or a *cipher*, being placed on the right. The zero 0 has no value, it is used only to
occupy a place, when there is nothing.else to be put in that
place.

Numbers expressed with the Roman Letters.

One	I	Twenty-five	XXV
Two	II	Twenty-six	XXVI
Three	III	Twenty-seven	XXVII
Four	*IIII	Twenty-eight	XXVIII
Five	V	Twenty-nine	*XXVIIII
Six	VI	Thirty	XXX
Seven	VII	Thirty-one	XXXI
Eight	VIII	Thirty-two	XXXII,&c.
Nine	*VIIII	Forty	*XXXX
Ten	X	Fifty	L
Eleven	XI	Sixty	LX
Twelve	XII	Seventy	LXX
Thirteen	XIII	Eighty	LXXX
Fourteen	*XIIII	Ninety	*LXXXX
Fifteen	XV	One hundred	C
Sixteen	XVI	Two hundred	CC
Seventeen	XVII	Three hundred	CCC
Eighteen	XVIII	Four hundred	CCCC
Nineteen	*XVIIII	Five hundred	D
Twenty	XX	Six hundred	DC
Twenty-one	XXI	Seven hundred	DCC
Twenty-two ·	XXII	Eight hundred	DCCC
Twenty-three	XXIII	Nine hundred	DCCCC
Twenty-four	*XXIIII	One thousand	M

One thousand, eight hundred, and twenty-six MDCCCXXVI

A man has a carriage worth seven hundred and sixty-eight dollars ;
and two horses, one worth two hundred and seventy-three dollars, and
the other worth two hundred and forty-seven dollars ; how many dollars are the whole worth ?

These numbers may be written as follows :—
    Operation.

DCCLXVIII dolls. ⎫   To add these numbers together it is easy
CCLXXIII dolls. ⎬ to see that it will be the most convenient to
CCXXXXVII dolls. ⎬ commence on the right, and count the Is
――――― ⎪ first. We find eight of them, which we
MCCLXXXVIII dolls. ⎭ should write thus VIII, but observing that

* It is usual to write four IV, instead of IIII, and nine IX, instead of VIIII,
and forty XL, instead of XXXX, and ninety XC, instead of LXXXX, &c. in
which a small character before a large, takes out its value from the large.
This is more convenient when no calculation is to be made. But when they
are to be used in calculation, the method given in the text is best.

Eleven is written thus, 11, with two 1s. The 1 on the left expresses *one ten ;* and the one on the right expresses *one unit,* or one added to ten. Twelve is written 12; the 1 on the left signifies one ten, and the 2 on the right signifies two units, and the whole is properly read *ten and two.*

there are more Vs we set down only III, reserving the V and counting it with the other Vs. Counting the Vs we find two, and the one which we reserved makes three. Three Vs are equivalent to one X and one V. We write the V and reserve the X. Counting the Xs, we find seven of them, and the one which was reserved makes eight. Eight Xs are equivalent to LXXX. We write the three Xs and reserve the L. Counting the Ls, we find two of them, and the one which was reserved makes three. Three Ls are equivalent to CL. We write the L and reserve the C. Counting the Cs, we find six of them, and the one which was reserved makes seven. Seven Cs are equivalent to DCC. We write the CC and reserve the D. Counting the Ds we find one, and the one which was reserved makes two. Two Ds are equivalent to M. The whole sum therefore is MCCLXXXVIII dollars.

The general rule for addition, therefore is, *to begin with the characters which express the lowest numbers and count all of each kind together without regard to their value, only observing that five Is make one V, and that two Vs make one X, and that five Xs make one L, &c., and setting them down accordingly.*

A man having one hundred and seventy-eight dollars, paid away seventy-nine dollars for a horse ; how many had he left?

Operation.

CLXXVIII dolls.   To perform this operation we begin at the
LXXVIIII dolls. right hand, and take the Is from the Is, the
  Vs from the Vs, &c. But a difficulty imme-
LXXXXVIII dolls. diately occurs, for we cannot take IIII from III ; it is necessary therefore to take the IIII from VIII, that is, from IIIIIIII, which leaves IIII ; these we set down. Since we have used the V in the upper line, it will be necessary to take the V in the lower line from one of the Xs, that is from VV. V from VV, leaves V, which we set down. Having used one of the Xs, there is but one left. We cannot take XX from X, we must therefore use the L, which is equivalent to five Xs, which, added to the one X, make XXXXXX ; from these we take XX and there remain XXXX, which we set down. Since the L in the upper line is already used, it is necessary to take the L in the lower line from the C which is equivalent to LL ; one L taken from these, leaves L, which we set down. The whole remainder therefore is LXXXXVIII dolls.

Hence the general rule for taking one number from another, expressed by the Roman characters, is, *to begin with the characters expressing the lowest numbers, and take those of the same kind from each other, when practicable, but if any of the numbers to be subtracted exceed those from which they are to be taken, a character of the next highest order must be taken, and reduced to the order required, and joined with the others from which the subtraction is to be made.*

This process is called subtraction.

The following is the manner of writing the numbers from nine to ninety-nine, inclusive.

The first column contains the figures, the second shows the proper mode of expressing them in words and the way in which they are always to be understood, and the third contains the names which are commonly applied. The common names are expressive of their signification, but not so much so as those in the second column.

Figures.	Proper mode of expressing them in words.	Common Names.
10.	One Ten or simply Ten.	Ten.
11.	Ten and one.	Eleven.
12.	Ten and two.	Twelve.
13.	Ten and three.	Thirteen.
14.	Ten and four.	Fourteen.
15.	Ten and five.	Fifteen.
16.	Ten and six.	Sixteen.
17.	Ten and seven.	Seventeen.
18.	Ten and eight.	Eighteen.
19.	Ten and nine.	Nineteen.
20.	Two tens.	Twenty.
21.	Two tens and one.	Twenty-one.
22.	Two tens and two.	Twenty-two.
23.	Two tens and three.	Twenty-three.
24.	Two tens and four.	Twenty-four.
25.	Two tens and five.	Twenty-five.
26.	Two tens and six.	Twenty-six.
27.	Two tens and seven.	Twenty-seven.
28.	Two tens and eight.	Twenty-eight.
29.	Two tens and nine.	Twenty-nine.
30.	Three tens.	Thirty.
31.	Three tens and one.	Thirty-one.
32, &c.	Three tens and two.	Thirty-two.
40.	Four tens.	Forty.
41, &c.	Four tens and one.	Forty-one.
50.	Five tens.	Fifty.
51, &c.	Five tens and one.	Fifty-one.
60.	Six tens.	Sixty.
61, &c.	Six tens and one.	Sixty-one.
70.	Seven tens.	Seventy.
71, &c.	Seven tens and one.	Seventy-one.
80.	Eight tens.	Eighty
81, &c.	Eight tens and one.	Eighty-one.

Figures	*Proper mode of expressing them in words.*	*Common Names.*
90.	Nine tens.	Ninety.
91, &c.	Nine tens and one.	Ninety-one.
99.	Nine tens and nine.	Ninety-nine.

Nine tens and nine or ninety-nine is the largest number that can be expressed by two figures. If one be added to nine tens and nine, it makes *ten tens*, or *one hundred.* To express one hundred we use the first figure again; but in order to show that it has a new value, it is put in another place, which is called the *hundreds' place.* The hundreds' place is the third place counting from the right. One hundred is written, 100; two hundred is written, 200; three hundred is written, 300. The zeros on the right have no value; their only purpose is to occupy the two first places, so that the figures 1, 2, 3, &c. may stand in the third place.

The figures in the second place, we observe, have the same value whether the first place be occupied by a zero or by a figure: for example, in 20 and in 23 the 2 has precisely the same value; it is two tens or twenty in both. In the first there is nothing added to the twenty, and in the second three is added to it.

It is the same with figures in the third place. They have the same value, whether the two first places are occupied by zeros or figures. In 400, 403, 420, and 435, the 4 has the same value in each, that is four hundred. The value of every figure, therefore, depends upon its place as counted from the right towards the left. A figure standing in the first place signifies so many units; the same figure standing in the second place signifies so many tens; and the same figure standing in the third place signifies so many hundreds. For example, 333, the three on the right signifies three units, the three in the second place signifies three tens or thirty, and the 3 in the third place signifies three hundreds. The number is read three hundreds, three tens, and three, or three hundred and thirty-three. We have seen that all the numbers from ten to twenty, from twenty to thirty, &c. are expressed by adding units to the tens; in the same manner all the numbers from one hundred to two hundred, from two hundred to three hundred, &c. are expressed by adding tens and units to the hundreds.—For example, to express five hundred and eighty-two, we write five hundreds, eight tens, and two units thus, 582.

The l looking over the above examples it will be observed,
is 999, the three first places on the right have distinct names,
hundredunits, tens, hundreds; and that the three next places are
we halled *thousands*, the first being called simply thousands;
*thousan*oond, tens of thousands; the third, hundreds of thou-
show th   In the same manner there are three places appro-
ther to d to millions, and distinguished in the same way, viz.
thousands, tens of millions, hundreds of millions. The same
which is of all the other names, three places being appropriat-
pressed bch name. From this circumstance it is usual to di-
sands.   figures into periods of three figures each. This
It is easyery much facilitates the reading and writing of
may be con;bers. Indeed it enables us to read a number con-
moved one rany number of figures, as easily as we can read
and since nces. This is illustrated in the following example.
use, there c
which cannot
We sometim,
hand place, un.
place, or the co
those in the third place, or the collection of hundreds, *units*
of the third order, &c.

Quadrillions Trillions Billions Millions Thousands Units

The following table exhibits the first nine places or orders,
With their names, and contains a few examples to illustrate
them.

Figures	*Proper mode of expressing them in words.*	*Common.*
90.	Nine tens.	Ninety.
91, &c.	Nine tens and one.	Ninety-o[ne.]
99.	Nine tens and nine.	Ninety-n[ine.]

Nine tens and nine or ninety-nine is the largest [number] that can be expressed by two figures. If one be [added to] nine tens and nine, it makes *ten tens*, or *one hund*[red. To] express one hundred we use the first figure agai[n; in] order to show that it has a new value, it is put [in a new] place, which is called the *hundreds' place.* The [hundreds'] place is the third place counting from the right. [One hun-] dred is written, 100; two hundred is written, [200; three] hundred is written, 300. The zeros on the r[ight have no] value; their only purpose is to occupy the two [places,] so that the figures 1, 2, 3, &c. may stand in th[e third.]

The figures in the second place, we obs[erve, have the] same value whether the first place be occupi[ed or not] by a figure: for example, in 20 and in 23 t[he 2 has equal-] ly the same value; it is two tens or twent[y.]

Words						3rd or hundreds' place.	2d or tens' place.	1st or units' place.
								7
							3	0
							4	6
						8	0	0
						7	0	3
						5	4	0
						6	5	8
					6	0	0	0
Six t[ens...]					6	0	0	5
Six thousands and four tens, or six thousand and forty					6	0	4	0
Six thousands and four tens and five units, or six thousand and forty-five					6	0	4	5
Six thousands and seven hundreds					6	7	0	0
Six thousand, seven hundred, and five					6	7	0	5
Six thousand, seven hundred, and forty					6	7	4	0
Six thousand, seven hundred, and forty-five					6	7	4	5
Four tens of thousands, or forty thousand				4	0	0	0	0
Forty thousand and three				4	0	0	0	3
Forty thousand, five hundred and three				4	0	5	0	3
Forty-seven thousand, five hundred, and eighty three				4	7	5	8	3
Four hundred and twenty-six thousand, eight hundred and fifty-three			4	2	6	8	5	3
Three hundred and twenty-eight millions, four hundred and thirty-five thousand, six hundred and eighty-seven	3	2	8	4	3	5	6	8	7
Three hundred millions	3	0	0	0	0	0	0	0	0
Twenty millions		2	0	0	0	0	0	0	0
Eight millions			8	0	0	0	0	0	0
Four hundred thousand				4	0	0	0	0	0
Thirty thousand					3	0	0	0	0
Five thousand						5	0	0	0
Six hundred							6	0	0
Eighty								8	0
Seven									7

In looking over the above examples it will be observed, that the three first places on the right have distinct names, viz. units, tens, hundreds; and that the three next places are all called *thousands*, the first being called simply thousands; the second, tens of thousands; the third, hundreds of thousands. In the same manner there are three places appropriated to millions, and distinguished in the same way, viz. millions, tens of millions, hundreds of millions. The same is true of all the other names, three places being appropriated to each name. From this circumstance it is usual to divide the figures into periods of three figures each. This division very much facilitates the reading and writing of large numbers. Indeed it enables us to read a number consisting of any number of figures, as easily as we can read three figures. This is illustrated in the following example.

Quintillions	Quadrillions.	Trillions	Billions	Millions	Thousands.	Units
Hundreds Tens Units	Hundreds Tens Units	Hundreds Tens Units	Hundreds Tens Units	Hundreds Tens Units	Hundreds Tens Units	Hundreds Tens Units
3 8 5,	6 7 9,	2 5 8,	6 7 3,	4 6 2,	9 2 7,	6 4 8

We have only to make ourselves familiar with reading and writing the figures of one period, and we shall then be able to read or write as many periods as we please, if we know the names of the periods.

It is to be observed that the unit of the first period is simply one; the unit of the second period is a collection of a thousand simple units; the unit of the third period is a collection of a thousand units of the second period, or a million of simple units; and so on as we proceed towards the left, each period contains a thousand units of the period next preceding it.

The figures of each period are to be read in precisely the same manner as the figures of the right hand period. At the end of each period, except the right hand period, the name of the period is to be pronounced. The right hand period is always understood to be units without mention being made of the name.

In the above example, the right hand period is read, six hundred and forty-eight (*units* being understood.) The second period is read in the same manner, nine hundred and twenty-seven,—but here we must mention the name of the period at the end ; we say, therefore, nine hundred and twenty-seven *thousand.* If we would put the two periods together, we begin on the left and say, nine hundred and twenty-seven thousand, six hundred and forty-eight. The third period is read four hundred and sixty-two,—adding the name of the period, it becomes four hundred and sixty-two *millions :* and the three periods are read together, four hundred and sixty-two *millions,* nine hundred and twenty-seven *thousand,* six hundred and forty-eight.

Beginning at the left hand of the above example, the several periods are read separately as follows—three hundred and eighty-five ; six hundred and seventy-nine ; two hundred and fifty-eight ; six hundred and seventy-three ; four hundred and sixty-two ; nine hundred and twenty-seven ; six hundred and forty-eight. Giving each period its name and putting all together as one number, it becomes three hundred and eighty-five *quintillions ;* six hundred and seventy-nine *quadrillions ;* two hundred and fifty-eight *trillions ;* six hundred and seventy-three *billions ;* four hundred and sixty-two *millions ;* nine hundred and twenty-seven *thousand ;* six hundred and forty-eight.

The names of the periods are derived from the Latin numerals, by giving them the termination *illion* and making some other alterations, so as to render the pronunciation easy. After quintillions come *sextillions, septillions, octillions, nonillions, decillions, undecillions, duodecillions, &c.*

A number dictated or enunciated, is written by beginning at the left hand, and proceeding towards the right, care being taken to give each figure its proper place. If any place is omitted in the enunciation, the place must be supplied with a zero. If, for example, the number were three hundred and twenty-seven thousand, and fifty-three ; we observe that the highest period mentioned is thousands, which is the second period, and that there are hundreds mentioned in this period, (that is, hundreds of thousands,) this period is therefore filled, and the number will consist of six places. We first write 3 for the three hundred thousand, then 2 immediately after it for the twenty thousand, then 7 for the seven thousand ; there were no hundreds mentioned in the enunciation, we must put a *zero* in the hundreds' place, then 5 for the tens, and 3 for the units, and the number will stand thus, 327,053.

Let the number be *fifty-three millions, forty thousand, six hundred and eight*. Millions is the third period, and tens of millions is the highest place mentioned, hence there will be but two places occupied in the period of millions, and the whole number will consist of eight places. We first write 53 for the millions. In the period of thousands there is only one place mentioned, that is, tens of thousands, we must put a *zero* in the hundreds of thousands' place, then 4 for the forty thousand, then a *zero* again in the thousands' place ; in the next period we write 6 for the six hundred, there being no tens in the example we put a *zero* in the tens' place, and then 8 for the eight units, and the whole number will stand thus, 53,040,608.

Whole periods may sometimes be left out in the enunciation. When this is the case, the places must be supplied by zeros. Great care must be taken in writing numbers, to use precisely the right number of places, for if a mistake of a single place be made, all the figures at the left of the mistake, will be increased or diminished tenfold.*

ADDITION.

II. We have seen how numbers are formed by the successive addition of units. It often happens that we wish to put together two or more numbers, and ascertain what number they will form.

A person bought an orange for 5 cents, and a pear for 3 cents ; how many cents did he pay for both ?

* The custom of using nine characters, and consequently the tenfold ratio of the places, is entirely arbitrary; any other number of figures might be used by giving the places a ratio corresponding to the number of figures, if we had only the seven first figures for example, the ratio of the places would be eight fold, and we should write numbers, in every other respect, as we do now. It would be necessary to reject the names eight and nine, and use the name of ten for eight. Twenty would correspond to the present sixteen : and one hundred, to the present sixty-four, &c. The following is an example of the eight fold ratio, with the numbers of the ten fold ratio corresponding to them.

To answer this question it is necessary to put together the numbers 5 and 3. It is evident that the first time a child undertakes to do this, he must take one of the numbers, as 5, and join the other to it a single unit at a time, thus 5 and 1 are 6, 6 and 1 are 7, 7 and 1 are 8; 8 is the *sum* of 5 and 3. A child is obliged to go through the process of adding by units every time he has occasion to put two numbers together, until he can remember the results. This however he soon learns to do if he has frequent occasion to put numbers together. Then he will say directly that 5 and 3 are 8, 7 and 4 are 11, &c.

Before much progress can be made in arithmetic, it is necessary to remember the sums of all the numbers from one to ten, taken two by two in every possible manner. These are all that are absolutely necessary to be remembered. For when the numbers exceed ten, they are divided into two or more parts and expressed by two or more figures, neither of which can exceed nine. This will be illustrated by the examples which follow.

A man bought a coat for twenty-four dollars, and a hat for eight dollars. How much did they both come to?

Operation.

Coat 24 dolls.   In this example we have 8 dolls. to
Hat   8 dolls.   add to 24 dolls. Here are twenty dolls.
  —   and four dolls. and eight dolls.   Eight
Both 32 dolls.   and four are twelve, which are to be join-

Eight fold				Ten fold	Eight fold		Ten fold
One	1	corresp. to		1	Fifteen	15 corresp. to	13
Two	2	-	-	2	Sixteen	16 - -	14
Three	3	•		3	Seventeen	17 - - -	15
Four	4	-		4	Twenty	20 - -	16
Five	5	-	-	5	Thirty	30 - - -	24
Six	6	-	- -	6	Forty	40 - . -	32
Seven	7	-	- - -	7	Fifty	50 - - -	40
Ten	10	-	- -	8	Sixty	60 - -	48
Eleven	11	-	- - -	9	Seventy	70 - - -	56
Twelve	12	-	- -	10	One hundred	100, &c. -	64
Thirteen	13	-	- - -	11	One thousand	1000 - -	512
Fourteen	14	-	- -	12			

In the same manner if we had twelve figures, the places would have been in a thirteen fold ratio.

The ten fold ratio was probably suggested by counting the fingers. This is the most convenient ratio. If the ratio were less, it would require a larger number of places to express large numbers. If the ratio were larger, it would not require so many places indeed, but it would not be so easy to perform the operations as at present on account of *the numbers in each place being so large.*

ed to twenty. But twelve is the same as ten. and two, therefore we may say twenty and ten are thirty and two are thirty-two.

A man bought a cow for 27 dolls. and a horse for 68 dolls. How much did he give for both ?

Operation.

Cow 27 dolls.   In this example it is proposed to add
Horse 68 dolls.   together 27 and 68. Now 27 is 2 tens
      —   and 7 units ; and 68 is 6 tens and 8
Both 95 dolls.   units. 6 tens and 2 tens are 8 tens ;
and 8 units and 7 units are 15, which is 1 ten and 5 units ; this joined to 8 tens makes 9 tens and 5 units, or 95.

A man bought ten barrels of cider for 35 dolls., and 7 barrels of flour for 42 dolls., a hogshead of molasses for 36 dolls., a chest of tea for 87 dolls., and 3 hundred weight of sugar for 24 dolls. What did the whole amount to ?

Operation.

Cider 35 dolls.   In this example there are five numbers
Flour 42 dolls.   bers to be added together. We ob-
Molasses 36 dolls.   serve that each of these numbers con-
Tea 87 dolls.   sists of two figures. It will be most
Sugar 24 dolls.   convenient to add together either all
      —   the units, or all the tens first, and then
Amount 224 dolls. the other. Let us begin with the
tens. 3 tens and 4 tens are seven tens, and 3 are 10 tens, and 8 are 18 tens, and 2 are 20 tens, or 200. Then adding the units, 5 and 2 are 7, and 6 are 13, and 7 are 20, and 4 are 24, that is, 2 tens and 4 units ; this joined to 200 makes 224.

It would be still more convenient to begin with the units, in the following manner ; 5 and 2 are 7, and 6 are 13, and 7 are 20, and 4 are 24, that is 2 tens and 4 units ; we may now set down the 4 units, and reserving the 2 tens add them with the other tens, thus : 2 tens (which we reserved) and 3 tens are 5 tens, and 4 are 9 tens, and 3 are 12 tens, and 8 are 20 tens, and 2 are 22 tens, which written with the 4 units make 224 as before.

A general has three regiments under his command ; in the first there are 478 men ; in the second 564 ; and in the third 593. How many men are there in the whole ?

Operation.

First reg.	478 men
Second reg.	564 men
Third reg.	593 men
	———
In all	1,635 men

In this example, each of the numbers is divided into three parts, hundreds, tens, and units. To add these together it is most convenient to begin with the units as follows; 8 and 4 are 12, and 3 are 15, that is, 1 ten and 5 units. We write down the 5 units, and reserving the 1 ten, add it with the tens. 1 ten (which we reserved) and 7 tens are 8 tens, and 6 are 14 tens, and 9 are 23 tens, that is, 2 hundreds and 3 tens. We write down the 3 tens, and reserving the 2 hundreds add them with the hundreds. 2 hundreds (which we reserved) and 4 hundreds are 6 hundreds, and 5 are 11 hundreds, and 5 are 16 hundreds, that is, 1 thousand and 6 hundreds. We write down the 6 hundreds in the hundreds' place, and the 1 thousand in the thousands' place.

The reserving of the tens, hundreds, &c. and adding them with the other tens, hundreds, &c. is called *carrying*. The principle of carrying is more fully illustrated in the following example.

A merchant had all his money in bills of the following description, one-dollar bills, ten-dollar bills, hundred-dollar bills, thousand-dollar bills, &c. each kind he kept in a separate box. Another merchant presented three notes for payment, one 2,673 dollars, another 849 dollars, and another 756 dollars. How much was the amount of all the notes; and how many bills of each sort did he pay, supposing he paid it with the least possible number of bills?

Operation.

Thous.	Hunds.	Tens.	Ones.
2	6	7	3
	8	4	9
	7	5	6
4	2	7	8

The first note would require 2 of the thousand-dollar bills; 6 of of the hundred-dollar bills; 7 ten-dollar bills; and 3 one-dollar bills. The second note would require 8 of the hundred-dollar bills; 4 ten-dollar bills; and 9 one-dollar bills. The third note would require 7 of the hundred-dollar bills; 5 ten-dollar bills; and 6 one-dollar bills. Counting the one-dollar bills, we find 18 of them. This may be paid with 1 ten-dollar bill and 8 one-dollar bills; putting this 1 ten-dollar bill with the other ten-dollar bills, we find 17 of them. This may be paid with 1 hundred-dollar bill, and 7 ten-dollar bills; putting this one-hundred dollar bill with the other hundred-dollar bills, we find 22 of them ; this may be paid with 2 of the thousand-dollar bills, and 2 of the hundred-dollar bills ; putting the 2 thousand-dollar bills with the other thousand-dollar bills, we find 4 of them. Hence the three notes may be paid with 4 of the thousand-dollar bills, 2 of the hundred-dollar bills, 7 ten-dollar bills, and 8 one-dollar bills, and the amount of the whole is 4,278 dollars.

Besides the figures, there are other signs used in arithmetic, which stand for words or sentences that frequently occur. These signs will be explained when there is occasion to use them.

A cross $+$ one mark being perpendicular, the other horizontal, is used to express, that one number is to be added to another. Two parallel horizontal lines $=$ are used to express equality between two numbers. This sign is generally read *is* or *are equal to*. Example $5 + 3 = 8$, is read 5 and 3 are 8; or 3 added to 5 is equal to 8 ; or 5 more 3 is equal to 8.; or more frequently 5 plus 3 is equal to 8 ; *plus* being the Latin word for *more*. These four expressions signify precisely the same thing.

Any number consisting of several figures may sometimes be conveniently expressed in parts by the above method. Example, $2358 = 2000 + 300 + 50 + 8 = 1000 + 1200 + 140 + 18$.

A man owns three farms, the first is worth 4,673 *dollars; the second,* 5,764 *dollars ; and the third,* 9,287 *dollars. How many dollars are they all worth?*

Perhaps the principle of carrying may be illustrated more plainly by separating the different orders of units from each other.

Operation.

4673 may be written	4000 +	600 +	70 +	3*
5764 - -	5000 +	700 +	60 +	4
9287 - -	9000 +	200 +	80 +	7

$$18000 + 1500 + 210 + 14$$

14
21 .
15 . .

Placing the results under each other, we have

18 . . . .

           18,000
         + 1,500
         +   210
19,724   +   14

$$= 19,724$$

In this example the sum of the units is 14, the sum of the tens is 21 tens or 210, the sum of the hundreds is 15 hundreds or 1,500, the sum of the thousands is 18 thousands or 18,000 ; these numbers being put together make 19,724.

If we take this example and perform it by carrying the tens, the same result will be obtained, and it will be perceived that the only difference in the two methods is, that in this, we add the tens in their proper places as we proceed, and in the other, we put it off until we have added each column, and then add them in precisely the same places.

Operation.

    4,673   Here as before the sum of the units is 14,
+5,764   but instead of writing 14 we write only the 4,
+9,287   and reserving the 1 ten, we say 1 (ten, which
            we reserved) and 7 are 8, and 6 are 14, and
=19,724   8 are 22 (tens) or 2 hundreds and 2 tens ;
setting down the 2 tens and reserving the hundreds, we say, 2 (hundreds, which we reserved) and 6 are 8, and 7 are 15, and 2 are 17 (hundreds) or 1 thousand and 7 hundreds ; writing down the 7 hundreds, and reserving the 1 thousand, we say, 1 (thousand, which we reserved) and 4 are 5, and 5 are 10, and 9 are 19 (thousands) or 1 ten-thousand and 9 thousands ; we write the 9 in its proper place, and since there is nothing more to add to the 1 (ten thousand) we write that down also, in its proper place. The answer is 19,724 dollars.

---

* It will be well for the learner to separate, in this way, several of the examples in Addition, because this method is frequently used for illustration in other parts of the book.

We may now observe another advantage peculiar to this method of notation. It is, that all large numbers are divided into parts, in order to express them by the different orders of units, and then we add each different order separately, and without regard to its name, observing only that ten, in an inferior order, is equal to one in the next superior order. By this means we add thousands, millions, or any of the higher orders as easily as we add units. If on the contrary we had as many names and characters, as there are numbers which we have occasion to use, the addition of large numbers would become extremely laborious. The other operations are as much facilitated as Addition, by this method of notation.

In the above examples the numbers to be added have been written under each other. This is not absolutely necessary; we may add them standing in any other manner, if we are careful to add units to units, tens to tens, &c., but it is generally most convenient to write them under each other, and we shall be less liable to make mistakes.

In the above examples we commenced adding the numbers at the top of each line, but it is easy to see that it will make no difference whether we begin at the top or bottom, since the result will be the same in either case.

Proof. The only method of proving addition, which can properly be called a proof, is by subtraction. This will be explained in its proper place.

The best way to ascertain whether the operation has been correctly performed, is to do it over again. But if we add the numbers the second time in the same order as at first, if a mistake has been made, we are very liable to make the same mistake again. To prevent this, it is better to add them in a reversed order, that is, if they were added downwards the first time, to add them upwards the second time, and *vice versa.*＊

＊ The method of omitting the upper line the second time, and then adding it to the sum of the rest is liable to the same objection, as that of adding the numbers twice in the same order, for it is in fact the same thing. If this method were to be used, it would be much better to omit the lower line instead of the upper one when they are added upward; and the upper line when added downward. This would change the order in which the numbers are put together.

The danger of making the same mistake is this: if in adding up a row of figures we should somewhere happen to say. 26 and 7 are 35, if we add it over again in the same way, we are very liable to say so again. But in adding it in another order it would be a very singular coincidence if a mistake of exactly the same number were made.

From what has been said it appears, that the operation of addition may be reduced to the following

RULE. *Write down the numbers in the most convenient manner, which is generally so that the units may stand under units, tens under tens, &c. First add together all the units, and if they do not exceed nine, write the result in the units' place; but if they amount to ten or more than ten, reserve the ten or tens, and write down the excess above even tens, in the units' place. Then add the tens, and add with them the tens which were reserved from the preceding column; reserve the tens as before, and set down the excess, and so on, till all the columns are added.*

MULTIPLICATION.

III. Questions often occur in addition in which a number is to be added to itself several times.

How much will 4 gallons of molasses come to at 34 cents a gallon?

　　34 cents

　　34 cents

　　34 cents

　　34 cents

　　——

Ans. 136 cents

This example may be performed very easily by the common method of addition. But it is easy to see that if it were required to find the price of 20, 30, or 100 gallons, the operation would become laborious on account of the number of times the number 34 must be written down.

I find in adding the units that 4 taken 4 times amounts to 16, I write the 6 and reserve the ten; 3 taken 4 times amounts to 12, and 1 which I reserved makes 13, which I write down, and the whole number is 136 cents.

If I have learned that 4 times 4 are 16, and that 4 times 3 are 12, it is plain that I need not write the number 34 but once, and then I may say that 4 times 4 are 16, reserving the ten and writing the 6 units as in addition. Then again 4 times 3 (tens) are 12 (tens) and 1 (ten which I reserved) are 13 (tens.)

Addition performed in this manner is called *Multiplication.* In this example 34 is the number to be *multiplied* or repeated, and 4 is the number by which it is to be multiplied; that is, it expresses the number of times 34 is to be taken.

The number to be multiplied is called the *multiplicand*, and the number which shows how many times the multiplicand is to be taken is called the *multiplier*. The answer or result is called the *product*. They are usually written in the following manner :

34 multiplicand

4 multiplier

———

136 product.

Having written them down, say 4 times 4 are 16, write the 6 and reserve the ten, then 4 times 3 are 12, and 1 (which was reserved) are 13.

In order to perform multiplication readily, it is necessary to retain in n emory the sum of each of the nine digits repeated from one to nine times ; that is, the products of each of the nine digits by themselves, and by each other. These are all that are absolutely necessary, but it is very convenient to remember the products of a much greater number. The annexed table, which is called the table of Pythagoras, contains the products of the first twenty numbers by the first ten.

TABLE OF PYTHAGORAS.

	2	3	4	5	6	7	8	9	10	11	12	13	14	15	16	17	18	19	20
1	2	3	4	5	6	7	8	9	10	11	12	13	14	15	16	17	18	19	20
2	4	6	8	10	12	14	16	18	20	22	24	26	28	30	32	34	36	38	40
3	6	9	12	15	18	21	24	27	30	33	36	39	42	45	48	51	54	57	60
4	8	12	16	20	24	28	32	36	40	44	48	52	56	60	64	68	72	76	80
5	10	15	20	25	30	35	40	45	50	55	60	65	70	75	80	85	90	95	100
6	12	18	24	30	36	42	48	54	60	66	72	78	84	90	96	102	108	114	120
7	14	21	28	35	42	49	56	63	70	77	84	91	98	105	112	119	126	133	140
8	16	24	32	40	48	56	64	72	80	88	96	104	112	120	128	136	144	152	160
9	18	27	24	45	54	63	72	81	90	99	108	117	126	135	144	153	162	171	180
10	20	30	40	50	60	70	80	90	100	110	120	130	140	150	160	170	180	190	200

To form this table, write the numbers 1, 2, 3, 4, &c. as far as you wish the table to extend, in a line horizontally. This is the first or upper row. To form the second row, add these numbers to 'hemselves, and write them in a row directly under the first. Thus 1 and 1 are 2; 2 and 2 are 4; 3 and 3 are 6; 4 and 4 are 8; &c. To form the third row, add the second row to the first, thus 2 and 1 are 3; 4 and 2 are 6; 6 and 3 are 9; 8 and 4 are 12; &c. This will evidently contain the first row three times. To form the fourth row, add the third to the first, and so on, till you have formed as many rows as you wish the table to contain.

When the formation of this table is well understood, the mode of using it may be easily conceived. If for instance the product of 7 by 5, that is, 5 times 7 were required, look for 7 in the upper row, then directly under it in the fifth row, you find 35, which is 7 repeated 5 times. In the same manner any other product may be found.

If you seek in the table of Pythagoras for the product of 5 by 7, or 7 times 5, look for 5 in the first row, and directly under it in the seventh row you will find 35, as before. It appears therefore that 5 times 7 is the same as 7 times 5. In the same manner 4 times 8 are 32, and 8 times 4 are 32; 3 times 9 are 27, and 9 times 3 are 27. In fact this will be found to be true with respect to all the numbers in the table. From this we should be led to suppose, that, whatever be the two numbers which are to be multiplied together, the product will be the same, whichsoever of them be made the multi-plier.

The few products contained in the table of Pythagoras are not sufficient to warrant this conclusion. For analogical reasoning is not allowed in mathematics, except to discover the probability of the existence of facts. But the facts are not to be admitted as truths until they are demonstrated. I shall therefore give a demonstration of the above fact; which, besides proving the fact, will be a good illustration of the manner in which the product of two numbers is formed.

There is an orchard, in which there are 4 rows of trees, and there are 7 trees in each row.

. . . . If one tree be taken from each
. . . . row, a row may be made consisting
. . . . of four trees; then one more taken
. . . . from each row will make another
row of four trees; and since there are seven trees in each

row, it is evident that in this way seven rows, of four trees
each, may be made of them. But the number of trees remains the same, which way soever they are counted.

Now whatever be the number of trees in each row, if they
are all alike, it is plain that as many rows, of four each, can
be made, as there are trees in a row. Or whatever be
the number of rows of seven each, it is evident that seven
rows can be made of them, each row consisting of a number
equal to the number of rows. In fine, whatever be the number of rows, and whatever be the number in each row, it is
plain that by taking one from each row a new row may be
made, containing a number of trees equal to the number of
rows, and that there will be as many rows of the latter kind,
as there were trees in a row of the former kind.

The same thing may be demonstrated abstractly as follows: 6 times 5 means 6 times each of the units in 5; but
6 times 1 is 6, and 6 times 5 will be 5 times as much, that
is, 5 times 6.

Generally, to multiply one number by another, is to repeat
the first number as many times as there are units in the
second number. To do this, each unit in the first must be
repeated as many times as there are units in the second.
But each unit of the first repeated so many times, makes a
number equal to the second; therefore the second number
will be repeated as many times as there are units in the first.
Hence the product of two numbers will always be the same,
whichsoever be made multiplier.

What will 254 pounds of meat cost, at 7 cents per pound?

This question will show the use of the above proposition;
for 254 pounds will cost 254 times as much as 1 pound; but
1 pound costs 7 cents, therefore it will cost 254 times 7.
But since we know that 254 times 7 is the same as 7 times
254, it will be much more convenient to multiply 254 by 7.
It is easy to show here that the result must be the same; for
254 pounds at 1 cent a pound would come to 254 cents; at
7 cents a pound therefore it must come to 7 times as much.

Operation.

254

7

———

Ans. 1778 cents.

Here say 7 times 4 are 28; reserving the 2 (tens) write the 8 (units);
then 7 times 5 (tens) are 35 (tens) and
2 (tens) which were reserved are 37
(tens); write the 7 (tens) and reserve the 3 (hundreds);

then 7 times 2 (hundreds) are 14 (hundreds) and 3 which were reserved are 17 (hundreds). The answer is 1778 cents ; and since 100 cents make a dollar, we may say 17 dollars and 78 cents.

The process of multiplication, by a single figure, may be expressed thus : *Multiply each figure of the multiplicand by the multiplier, beginning at the right hand, and carry as in addition.*

IV.   *What will 24 oxen come to, at 47 dollars apiece ?*

It does not appear so easy to multiply by 24 as by a num ber consisting of only one figure ; but we may first find the price of 6 oxen, and then 4 times as much will be the price of 24 oxen.

Operation.
47
6
———
282 dolls. price of 6 oxen.
4
———
1128 dolls. price of 24 oxen.

Or thus   47
4
———
188 dolls. price of 4 oxen.
6
———
1128 dolls. price of 24 oxen.

A number which is a product of two or more numbers is called a *composite* or *compound* number.   The numbers, which, being multiplied together, produce the number, are called *factors* of that number.   4 is a composite number, its factors are 2 and 2, because 2 times 2 are 4.   6 is also a composite number, its factors are 2 and 3.   The numbers 8, 9, 10, 12, 14, 15, &c. are composite numbers ; some of them have only two factors, and some have several.   The sign $\times$, a cross, in which neither of the marks is either horizontal or perpendicular, is used to express multiplication. Thus $3 \times 2 = 6$, signifies 2 times 3 are equal to 6. $2 \times 3 \times 5 = 30$, signifies 3 times 2 are 6, and 5 times 6 are 30.

Numbers which have several factors, may be divided into a number of factors, less than the whole number of factors, in several ways. 24, for example, has 4 factors, thus, $2 \times 2 \times 2 \times 3 = 24$. This may be divided into 2 factors and into 3 factors in several different ways. Thus $4 \times 6 = 24$; $2 \times 2 \times 6 = 24$; $3 \times 8 = 24$; $2 \times 12 = 24$; $2 \times 6 \times 2 = 24$.

When several numbers are to be multiplied together, it will make no difference in what order they are multiplied, the result will always be the same.

What will be the price of 5 loads of cider, each load containing 7 barrels, at 4 dollars a barrel?

Now 5 loads each containing 7 barrels, are 35 barrels. 35 barrels at 4 dollars a barrel, amount to 140 dollars. Or we may say one load comes to 28 dollars, and 5 loads will come to 140 dollars. Or lastly, 1 barrel from each load will come to 20 dollars, and 7 times 20 are 140.

	Thus	Or		Or	
	7		7		5
	5		4		4
	---		---		---
	35		28		20
	4		5		7
	---		---		---
	140		140		140

What is the price of 23 loads of hay, at 34 dolls. a load?

$$34$$
$$2$$
$$\overline{}$$

68 dolls. price of 2 loads.

$$34$$
$$7$$
$$\overline{}$$

238 dolls. price of 7 loads.
$$3$$
$$\overline{}$$

714 dolls. price of 21 loads.
$+$ 68 dolls. price of 2 loads.
$$\overline{}$$
$=$ 782 dolls. price of 23 loads.

Multiply 328 *by* 112.
$$112 = 4 \times 7 \times 4$$

                                     328
                                        4
                             ————
                             1312 product by 4
                                   7
                             ————
                             9184 product by 28
                                   4
                             ————
                           36736 product by 112

It is easy to see that we may multiply by any other number in the same manner.

This operation may be expressed as follows. To multiply by a composite number: *Find two or more numbers, which being multiplied together will produce the multiplier; multiply the multiplicand by one of these numbers, and then that product by another, and so on, until you have multiplied by all the factors, into which you had divided the multiplier, and the last product will be the product required.*

If the multiplier be not a composite number, or if it cannot be divided into convenient factors: *Find a composite number as near as possible to the multiplier, but smaller, and multiply by it according to the above rule, and then add as many times the multiplicand, as this number falls short of the multiplier.*

V. I have shown how to multiply any number by a single figure ; and when the multiplier consists of several figures, how to decompose it into such numbers as shall contain but one figure. It remains to show how to multiply by any number of figures ; for the above processes will not always be found convenient.

The most simple numbers consisting of more than one figure are 10, 100, 1000, &c. It will be very easy to multiply by these numbers, if we recollect that any figure written in the second place from the right signifies ten times as many as it does when it stands alone, and in the third place, one hundred times as many, and so on. If a zero be annexed at the right of a figure or any number of figures, it is evident that they will all be removed one place towards the left, and consequently become ten times as great ; if two zeros be annexed they will be removed two places, and will be one hundred times as great, &c. Hence, *to multiply by*

12

any number consisting of 1, *with any number of zeros at the*
right of it it is sufficient to annex the zeros to the multipli
cand.

$$1 \times 10 = 10 \qquad 1 \times 100 = 100$$
$$2 \times 10 = 20 \qquad 3 \times 100 = 300$$
$$3 \times 10 = 30 \qquad 5 \times 100 = 500$$
$$27 \times \quad 10 = \qquad 270$$
$$42 \times \quad 100 = \qquad 4200$$
$$368 \times 1000 = 368000$$

VI.   When the multiplier is 20, 30, 40, 200, 300, 2000,
4000, &c.   These are composite numbers, of which 10, or
100, or 1000, &c. is one of the factors.   Thus $20 = 2 \times$
10; $30 = 3 \times 10$; $300 = 3 \times 100$; &c.   In the same
manner $387000 = 387 \times 1000$.

How much will 30 *hogsheads of wine come to, at* 87 *dollars*
per hogshead ?

> *Operation.*
> 87
> 3
> ———
> 261 dolls. price of 3 hhds.
> 10
> ———
> 2610 dolls. price of 30 hhds.

More simply thus   87
        30
       ———
2610 dolls. price of 30 hhds.

•   It appears that it is sufficient in this example to multiply
by 3 and then annex a zero to the product.   If the number
of hogsheads had been 300, or 3000, two or three zeros must
have been annexed.   It is plain also that, *if there are zeros*
on the right of the multiplicand, they may be omitted until
the multiplication has been performed, and then annexed to
the product.

VII. *A man bought* 26 *pipes of wine, at* 143 *dollars a pipe; how much did they come to?*

$26 = 20 + 6.$　The operation may be performed thus:

$$143$$
$$6$$

858 dolls. price of 6 pipes

$$143$$
$$20$$

2860 dolls. price of 20 pipes
+　858 dolls. price of 6 pipes

= 3718 dolls. price of 26 pipes

The operation may be performed more simply thus;

$$143$$
$$26$$

2860 dolls. price of 20 pipes
+　858 dolls. price of 6 pipes

= 3718 dolls. price of 26 pipes

Or multiplying first by 6:

$$143$$
$$26$$

858 dolls. price of 6 pipes
+ 2860 dolls. price of 20 pipes

= 3718 dolls. price of 26 pipes

If the wages of 1 *man be* 438 *dollars for* 1 *year, what will be the wages of* 234 *men, at the same rate?*

Operation.
$$438$$
$$234$$

87600 dolls. wages of 200 men
+ 13140　do.　wages of 30 men
+　1752　do.　wages of 4 men

=102492 dolls. wages of 234 men

Or thus 438
      234
     ————

   1752 dolls. wages of 4 men
+ 13140 do. wages of 30 men
+ 87600 do. wages of 200 men
   ————

=102492 dolls. wages of 234 men

When we multiply by the 30 and the 200, we need not annex the zeros at all, if we are careful, when multiplying by the tens, to set the first figure of the product in the ten's place, and when multiplying by hundreds, to set the first figure in the hundred's place, &c.

Operation.
438
234
————
1752
1314.
876..
————
102,492

If we compare this operation with the last, we shall find that the figures stand precisely the same in the two.

We may show by another process of reasoning, that when we multiply units by tens, the first figure of the product should stand in the tens' place, &c. ; for units multiplied by tens ought to produce tens, and units multiplied by hundreds, ought to produce hundreds, in the same manner as tens multiplied by units produce tens.

If it take 853 dollars to support a family one year, how many dollars will it take to support 207 such families the same time ?

Operation.

  853   In this example I multiply first by the 7
  207 units, and write the result in its proper place ;
  —— then there being no tens, I multiply next by
 5971 the 2 hundreds, and write the first figure of
 1706 this product under the hundreds of the first
 ——   product ; and then add the results in the order
176571 in which they stand.

The general rule therefore for multiplying by any number of figures may be expressed thus : *Multiply each figure of the multiplicand by each figure of the multiplier separately, taking care when multiplying by units to make the first figure of the result stand in the units' place ; and when multiplying by tens, to make the first figure stand in the tens' place ; and when multiplying by hundreds, to make the first figure stand in the hundreds' place, &c.* and then add the several products together.

Note. It is generally the best way to set the first figure of each partial product directly under the figure by which you are multiplying.

Proof. The proper proof of multiplication is by division, consequently it cannot be explained here. There is also a method of proof by casting out the nines, as it is called. But the nature of this cannot be understood, until the pupil is acquainted with division. It will be explained in its proper place. The instructer, if he chooses, may explain the use of it here.

————

SUBTRACTION.

VIII. *A man having ten dollars, paid away three of them ; how many had he left ?*

We have seen that all numbers are formed by the successive addition of units, and that they may also be formed by adding together two or more numbers smaller than themselves, but all together containing the same number of units as the number to be formed. The number, 10 for example, may be formed by adding 3 to 7, $7 + 3 = 10$. It is easy to see therefore that any number may be decomposed into two or more numbers, which taken together, shall be equal to that number. Since $7 + 3 = 10$, it is evident that if 3 be taken from 10, there will remain 7.

The following examples, though apparently different, all require the same operation, as will be immediately perceived.

A man having 10 sheep sold 3 of them ; how many had he left ? That is, if 3 be taken from 10, what number will remain ?

12 *

A man gave 3 *dollars to one son, and* 10 *to another; how much more did he give to the one than to the other? That is, how much greater is the number* 10 *than the number* 3 ?

A man owing 10 *dollars, paid* 3 *dollars at one time, and the rest at another; how much did he pay the last time? That is, how much must be added to* 3 *to make* 10 ?

From Boston to Dedham it is 10 *miles, and from Boston to Roxbury it is only* 3 *miles; what is the difference in the two distances from Boston?*

A boy divided 10 *apples between two other boys; to one he gave* 3, *how many did he give to the other? That is, if* 10 *be divided into two parts so that one of the parts may be* 3, *what will the other part be?*

It is evident that the above five questions are all answered by taking 3 from 10, and finding the difference. This operation is called *subtraction*. It is the reverse of addition. *Addition* puts numbers together, *subtraction* separates a number into two parts.

A man paid 29 *dollars for a coat and* 7 *dollars for a hat, how much more did he pay for his coat than for his hat?*

In this example we have to take the 7 from the 29; we know from addition, that 7 and 2 are 9, and consequently that 22 and 7 are 29; it is evident therefore that if 7 be taken from 29 the remainder will be 22.

A man bought an ox for 47 *dollars; to pay for it he gave a cow worth* 23 *dollars, and the rest in money; how much money did he pay?*

Operation.

Ox 47 dollars.   Cow 23 dollars.

It will be best to perform this example by parts. It is plain that we must take the twenty from the forty, and the three from the seven; that is, the tens from the tens, and the units from the units. . I take twenty from forty, and there remains twenty. I then take three from seven, and there remains four, and the whole remainder is twenty-four. Ans. 24 dollars.

It is generally most convenient to write the numbers under each other. The smaller number is usually written under the larger. Since units are to be taken from units, and tens from .ens, it will be best to write units under units,

tens under tens, &c. as in addition. It is also most convenient, and, in fact, frequently necessary, to begin with the
units as in addition and multiplication.

Operation.

Ox   47 dollars.   I say first 3 from 7, and there will
Cow 23 dollars.   remain 4. Then 2 (tens) from 4
   —   (tens) and there will remain 2 (tens),
   24 difference.   and the whole remainder is 24.

*A man having 62 sheep in his flock, sold 17 of them ; how
many had he then ?*

Operation.

He had 62 sheep   In this example a difficulty immedi
Sold   17 sheep   ately presents itself, if we attempt to
   —   perform the operation as before ; for
Had left 45 sheep   we cannot take 7 from 2. We can,
however, take 7 from 62, and there remains 55 ; and 10
from 55, and there remains 45, which is the answer.

The same operation may be performed in another way,
which is generally more convenient. I first observe, that 62
is the same as 50 and 12 ; and 17 is the same as 10 and 7.
They may be written thus :

$62 = 50 + 12$ That is, I take one ten from the six
$17 = 10 + 7$ tens, and write it with the two units.
   ————————   But the 17 I separate simply into units
$45 = 40 + 5$ and tens as they stand. Now I can take
7 from 12, and there remains 5. Then 10 from 50, and there
remains 40, and these put together make 45.*
This separation may be made in the mind as well as to
write it down.

Operation.

   62   Here I suppose 1 ten taken from the 6 tens,
   17   and written with the 2, which makes 12. I say
   —   7 from 12, 5 remains, then setting down the 5, I
   45   say, 1 ten from 5 tens, or simply 1 from 5, and
there remains 4 (tens), which written down shows the remainder, 45.

The taking of the ten out of 6 tens and joining it with
the 2 units, is called *borrowing ten.*

* Let the *pupil perform* a large number of examples by separating
them in this way, when he first commences subtraction.

Sir Isaac Newton was born in the year 1642, *and he died in* 1727 ; *how old was he at the time of his decease ?*

It is evident that the difference between these two numbers must give his age.

Operation.

$$1600 + 120 + 7 = 1727$$
$$1600 + 40 + 2 = 1642$$

Ans. $80 + 5 = 85$ years old.

In this example I take 2 from 7 and there remains 5, which I write down. But since I cannot take 4 (tens) from 2 (tens,) I borrow 1 (hundred) or 10 tens from the 7 (hundreds,) which joined with 2 (tens) makes 12 (tens,) then 4 (tens) from 12 (tens) there remains 8 (tens,) which I write down. Then 6 (hundreds) from 6 (hundreds) there remains nothing. Also 1 (thousand) from 1 (thousand) nothing remains. The answer is 85 years.

A man bought a quantity of flour for 15,265 *dollars, and sold it again for* 23,007 *dollars, how much did he gain by the bargain ?*

Operation.

23,007   Here I take 5 from 7 and there remains
15,265   2 ; but it is impossible to take 6 (tens) from
———   0, and it does not immediately appear where
2   I shall borrow the 10 (tens,) since there is
nothing in the hundreds' place. This will be evident, however, if I decompose the numbers into parts.

Operation.

$$10,000 + 12,000 + 900 + 100 + 7 = 23,007$$
$$10,000 + 5,000 + 200 + 60 + 5 = 15,265$$
$$ 7,000 + 700 + 40 + 2 = 7,742$$

The 23,000 is equal to 10,000 and 13,000 ; this last is equal to 12,000 and 1,000 ; and 1,000 is equal to 900 and 100. Now I take 5 from 7, and there remains 2 ; 60 from 100, or 6 tens from 10 tens, and there remains 40, or 4 tens ; 2 hundreds from 9 hundreds, and there remains 7 hundreds ; 5 thousands from 12 thousands, and there remains 7 thousands ; and 1 ten-thousand from 1 ten-thousand, and nothing remains. The answer is 7,742 dollars.

This example may be performed in the same manner as

the others, without separating it into parts except in the mind.

I say 5 from 7, there remains 2 : then borrowing 10 (which must in fact come from the 3 (thousand), I say, 6 (tens) from 10' (tens) there remains 4 (tens ;) then I borrow ten again, but since I have already used one of these, I say, 2 (hundreds) from 9 (hundreds) there remains 7 (hundreds;) then I borrow ten again, and having borrowed one out of the 3 (thousand,) I say, 5 (thousand) from 12 (thousand) there remains 7 (thousand ;) then 1 (ten-thousand) from 1 (ten-thousand) nothing remains.   The answer is 7,742 as before.

The general rule for subtraction may be expressed thus : *The less number is always to be subtracted from the larger.   Begin at the right hand and take successively each figure of the less number from the corresponding figure of the larger number, that is, units from units, tens from tens, &c.   If it happens that any figure of the less number cannot be taken from the corresponding figure of the larger, borrow ten and join it with the figure from which the subtraction is to be made, and then subtract ; before the next figure is subtracted take care to diminish by one the figure from which the subtraction is to be made.*

N. B.   When two or more zeros intervene in the number from which the subtraction is to be made, all, except the first, must be called 9s in subtracting, that is, after having borrowed ten, it must be diminished by one, on account of the ten which was borrowed before.

Note.   It is usual to write the smaller number under the greater, so that units may stand under units, and tens under tens, &c.

Proof.   A man bought an ox and a cow for 73 dollars, and the price of the cow was 25 dollars ; what was the price of the ox?

The price of the ox is evidently what remains after taking 25 from 73.

Operation.

Ox and cow 73 dollars
Cow   25 do.
          —
Ox   48 do.

It appears that the ox cost 48 dollars.   If the cow cost 25 dollars, and the ox 48 dollars, it is evident that 25 and 48 added together must make 73 dollars, what they both cost.

Hence to prove subtraction, add the remainder and the smaller number together, and if the work is right their sum will be equal to the larger number.

Another method. If the ox cost 48 dollars, this number taken from 73, the price of both, must leave the price of the cow, that is, 25. Hence subtract the remainder from the larger number, and if the work is right, this last remainder will be equal to the smaller number.

Proof of addition. It is evident from what we have seen of subtraction, that when two numbers have been added together, if one of these numbers be subtracted from the sum, the remainder, if the work be right, must be equal to the other number. This will readily be seen by recurring to the last example. In the same manner if more than two numbers have been added together, and from the sum all the numbers but one, be subtracted, the remainder must be equal to that one.

DIVISION.

IX. *A boy having 32 apples wished to divide them equally among 8 of his companions; how many must he give them apiece ?*

If the boy were not accustomed to calculating, he would probably divide them, by giving one to each of the boys, and then another, and so on. But to give them one apiece would take 8 apples, and one apiece again would take 8 more, and so on. The question then is, to see how many times 8 may be taken from 32 ; or, which is the same thing, to see how many times 8 is contained in 32. It is contained four times. Ans. 4 each.

A boy having 32 apples was able to give 8 to each of his companions. How many companions had he ?

This question, though different from the other, we perceive, is to be performed exactly like it. That is, it is the question to see how many times 8 is contained in 32. We take away 8 for one boy, and then 8 for another, and so on.

A man having 54 cents, laid them all out for oranges, at 6 cents apiece. How many did he buy ?

It is evident that as many times as 6 cents can be taken from 54 cents, so many oranges he can buy.   Ans. 9 oranges

A man bought 9 *oranges for* 54 *cents; how much did he give apiece ?*

In this example we wish to divide the number 54 into 9 equal parts, in the same manner as in the first question we wished to divide 32 into 8 equal parts.   Let us observe, that if the oranges had been only one cent apiece, nine of them would come to 9 cents ; if they had been 2 cents apiece, they would come to twice nine cents ; if they had been 3 cents apiece, they would come to 3 times 9 cents, and so on. Hence the question is to see how many times 9 is contained in 54.   Ans. 6 cents apiece.

In all the above questions the purpose was to see how many times a small number is contained in a larger one, and they may be performed by subtraction.   If we examine them again we shall find also, that the question was, in the two first, to see what number 8 must be multiplied by, in order to produce 32 ; and in the third, to see what the number 6 must be multiplied by, to produce 54 ; in the fourth, to see what number 9 must be multiplied by, or rather what number must be multiplied by 9, in order to produce 54.

The operation by which questions of this kind are performed is called *division*.   In the last example, 54, which is the number to be divided, is called the *dividend*; 9, which is the number divided by, is called the *divisor ;* and 6, which is the number of times 9 is contained in 54, is called the *quotient*.

It is easy to see from the above reasoning, that the quotient and divisor multiplied together must produce the dividend ; for the question is to see how many times the divisor must be taken to make the dividend, or in other words to see what the divisor must be multiplied by to produce the dividend.   It is evident also, that if the dividend be divided by the quotient, it must produce the divisor.   For if 54 contains 6 nine times, it will contain 9 six times.

To prove division, multiply the divisor and quotient together, and if the work be right, the product will be the dividend.   Or divide the dividend by the quotient, and if the work be right, the result will be the divisor.

This also furnishes a proof for multiplication, for if the

quotient multiplied by the divisor produces the dividend, it is evident, that if the product of two numbers be divided by one of those numbers, the quotient must be the other num ber.

It appears that division is applied to two distinct purposes, though the operation is the same for both. The object of the first and fourth of the above examples is to divide the numbers into equal parts, and of the second and third to find how many times one number is contained in another. At present, we shall confine our attention to examples of the latter kind, viz. to find how many times one number is contained in another.

At 3 cents apiece, how many pears may be bought for 57 cents ?

It is evident, that as many pears may be bought, as there are 3 cents in 57 cents. But the solution of this question does not appear so easy as the last, on account of the greater number of times which the divisor is contained in the divi-, dend. If we separate 57 into two parts it will appear more easy.

$$57 = 30 + 27.$$

We know by the table of Pythagoras that 3 is contained in 30 ten times, and in 27 nine times, consequently it is contained in 57 nineteen times, and the answer is 19 pears.

How many barrels of cider, at 3 dollars a barrel, can be bought for 84 dollars ?

Operation.

$84 = 60 + 24$ 3 is contained in 6 twice, but in 6 tens it is contained ten times as often, or 20 times. 3 is contained in 24 eight times, consequently 3 is contained 28 times in 84. Ans. 28 barrels.

How many pence are there in 132 farthings ?

As many times as 4 farthings are contained in 132 far-things, so many pence there are.

Operation.

$132 = 120 + 12$ 120 is 12 tens, 4 is contained in 12 three times, consequently it is contained 30 times in 12 tens. 4 is contained 3 times in 12 units, consequently in 132 it is contained 33 times. Ans. 33 pence.

How many barrels of flour, at 5 dollars a barrel, may be bought for 785 dollars.

> ### Operation.
> 785 = 500 + 250 + 35

5 is contained in 5 once, and in 500 one hundred times. 250 is 25 tens. 5 is contained 5 times in 25, consequently 50 times in 250. 5 is contained 7 times in 35 units. In 785, 5 is contained 157 times. Ans. 157 barrels.

How many dollars are there in 7464 shillings ?

As many times as 6 shillings are contained in 7464 shillings, so many dollars there are.

> ### Operation.
> 7464 = 6000 + 1200 + 240 + 24

6 is contained 1000 times in 6000, 200 times in 1200, 40 times in 240, and 4 times in 24, making in all 1244 times.* Ans. 1244 dollars.

It is not always convenient to resolve the number into parts in this manner at first, but we may do it as we perform the operation.

• *In 126 days how many weeks ?*

> ### Operation.
> 126 = 70 + 56   Instead of resolving it in this man-

ner, we will write it down as follows.

> Dividend 126 (7 Divisor
> 70 —
> — 10
> 56   8
> 56 —
> — 18 quotient

I observe that 7 cannot be contained 100 times in 126, I therefore call the two first figures on the left 12 tens or 120, rejecting the 6 for the present. 7 is contained more than once and not so much as twice in 12, consequently in 12 tens it is contained more than 10 and less than 20 times. I take 10 times 7 or 70 out of 126, and there remains 56. Then 7 is contained 8 times in 56, and 18 times in 126. Ans. 18 weeks.

---

* Let the pupil perform a large number of examples in this manner when he first commences ; as he is obliged to separate the numbers into parts, he will at length come to the common method.

13

In 3756 pence how many four-pences ?

It is evident that this answer will be obtained by finding how many times 4 pence is contained in 3756 pence.

If we would solve this, as we did the first examples, it will stand thus :

$$3756 = 3600 + 120 + 36$$

But if we resolve it into parts, as we perform the operation, it will be done as follows :

```
Dividend 3756 (4 divisor
         3600 ——
         ——— 900 =· number that 4 is contained in 3600
          156  30    do.  -   -   -   -    120
          120   9    do.  -   -   -   -     36  ·
          ——  ——
           36 939    do.  -   -·  -   -   3756
           36
           ——
           · ·
```

Here I take the 37 hundreds alone, and see how many times 4 is contained in it, which I find 9 times, and since it is 37 hundreds, it must be contained 900 times. 900 times 4 is 3600, which I subtract from 3756, and there remains 156. It is now the question to find how many times 4 is contained in this. I take the 15 tens, rejecting the 6, and see how many times 4 is contained in it. It is contained 3 times, and since it is 15 tens, this must be 3 tens or 30 times. 30 times 4 is 120. This I subtract from 156, and there remains 36. 4 is contained in 36, 9 times ; hence it is contained in the whole 939 times. Ans. 939 four-pences.

If these partial numbers, viz. 3600, 120, and 36, are compared with the resolution of the number above, they will be found to be the same.

This operation may be abridged still more.

```
3756 (4
  36  ——
  ——  939 quotient.
  15
  12
  ——
  36
  36
  ——
  · ·
```

In this I say, 4.into 37 9 times, and set down the 9 in the quotient, without regarding whether it is hundreds, or tens, or units, but by the time I have' done dividing, if I set the other figures by the side of it, it will be brought into its proper place.   Then I say 9 times 4 are 36, and set it under the 37, as before, but do not write the. zeros by the side of it.   I then subtract 36 from 37, and there remains 1.   This of course is 100, but I do not mind it.   I then bring down the 5 by the side of the 1, which makes 15, or rather 150, but I call it 15.   Then I say, 4 into 15, 3 times, (this is 30, but I write only the 3 ;) I write the 3 by the side of the 9. Then I say, 3 times 4 is 12, which I write under the 15, and subtract it from 15, and there remains 3 (which is in fact 30.)   By the side of 3 I bring down the 6, which makes 36.   Then I say 4 into 36, 9 times, which I write in the quotient, by the side of the 93, and it. makes 939. The first 9 is now in the hundreds' place, and the 3 in the ten's place, as they ought to be.   If this operation be compared with the last, it will be found in substance exactly like it.   All the difference is, that in the last the figures are set down only when they are to be used.

A man employed a number of workmen, and gave them 27 dollars a month each ; at the expiration of one month, it took 10,125 dollars to pay them.   How many men were there ?

It is evident that to find the number of men we must find how many times 27 dollars is contained in 10,125 dollars.

This may be done in the same manner as we did the last, though it is attended with rather more difficulty, because the divisor consists of two figures.

Operation.

Dividend 10,125 (27 divisor
        8,100 ——
       ——— 300 = the number of times 27 is contained
       2,025   in 8,100
       1,890 70   do.   -   -   1,890
       ——— 5   do.   -   -   135
        135 ——
        135 375   do.   -   -   - 10,125

Common way.
10,125 (27
81
——   375 quotient.
202
. 189

135
135

. . .

I observe that there are not so many as 27 thousands, so I
conclude that the divisor is not contained 1000 times in the
dividend; I therefore take the three left hand figures, neg-
lecting the other two for the present. The three first are
101; (properly 10,100, but I notice only 101;) I seek how
many times 27 is contained in 101, and find between 3 and
4 times. I put 3 in the quotient, which, when the work is
done, must be 3 hundred, because 101 is 101 hundreds, but
disregarding this circumstance, I find how much 3 times 27
is, and write it under 101. 3 times 27 are 81; this subtracted from 101, leaves 20. By the side of 20 I bring down
2, the next figure of the dividend which was not used. This
makes 202, for the next partial dividend. I seek how many
times 27 is contained in this. I find 7 times. I write 7 in
the quotient. 7 times 27 are 189, which I subtract from
202, and find a remainder 13. By the side of 13 I bring
down 5, the other figure of the dividend, which makes 135
for the last partial dividend. I find 27 is contained 5 times
in this. I write 5 in the quotient. 5 times 27 are 135.
There is no remainder, therefore the division is completed.
Ans. 375 men.

The operation in the above example is precisely the same,
as in those which precede it; but it is more difficult to discover how many times the divisor is contained in the partial
dividends. When the divisor is still larger, the difficulty is
increased. I shall next show how this difficulty may be ob-
viated.

*In 31,755 days, how many years, allowing 365 days to the
year ?*

It is evident, that as many times as 365 is contained in
31,755, so many years there will be.

Operation.

Dividend 31755 (365 divisor
   .2920   ——
   ——   . 87 quotient.
   2555
   2555

   ——
   . . . .

I observe that 365 cannot be contained in 317, therefore I must take the four left hand figures, viz. 3175. In order to discover how many times 365 is contained in this, I observe, that 365 is more than 300, and less than 400. I say 300 is contained in 3100, or simply 3 is contained in 31, 10 times, but 365 being greater than 300, cannot be contained in it more than 9 times. Indeed if it were contained more than 9 times, it must have been contained in 317, which is impossible. 400 is contained in 3100, (or 4 in 31) 7 times. This is the limit the other way, for 365 being less than 400, must be contained at least as many times. It is contained therefore 7, or 8, or 9 times. The most probable are 8 and 9. I try 9. But instead of multiplying the whole number 365 by 9, I say 9 times 300 are 2700, or simply 9 times 3 are 27; then subtracting 2700 from 3170, there remains 470; I then say, 9 times 60 is 540, or simply 9 times 6 is 54, which being larger than 470, or 47, shows that the divisor is not contained 9 times. I next try 8 times, and say as before, 8 times 300 are 2400, which subtracted from 3170, leaves 770, then 8 times 60 are 480, which not being so large as 770, shows that the divisor is contained 8 times. I multiply the whole divisor by 8 (which is in fact 80,) the product is 2920. This subtracted from 3175 leaves 255. I then bring down the other 5, which makes the next partial dividend 2555. Now trying as before, I find that 3 is contained 8 times in 25, and 4 is contained 6 times. The limits are 6 and 8. It is probable that 7 is right. I multiply 365 by 7, and it makes 2555, which is exactly the number that I want. If I had wished to try 8, I should have said 8 times 3 are 24, which taken from 25 leaves 1. Then supposing 1 to be placed before the next figure, which is 5, it makes 15. 6 is not contained 8 times in 15, therefore 365 cannot be contained 8 times in 2555. The answer is 87 years.

The method of trying the first figure of the divisor into the *first figure,* or the first two figures of the partial dividend, generally enables us to tell what the quotient figure must be, within two or three, and it will always furnish the limits. Then if we try the second figure, we shall always make the limits smaller ; if any doubt then remains, which will not often be the case, we may try the third, and so on.

Divide 436940074 *by* 64237.

Operation.

Dividend 436940074 (64237 divisor.
　　　　385422*　　　 ———

　　　　 ———　　　 6802　quotient.

　　　 . 515180
　　　 . 513896*

　　　 ... 128474
　　　 ... 128474*

Proof　436940074

In this example I seek how many times 6, the first figure of the divisor, is contained in 43, the first two figures on the left of the dividend ; I find 7 times, and 7 is contained 6 times. The limits are 6 and 7. 7 times 6 are 42, and 42 from 43 leaves 1, which I suppose placed by the side of 6 ; this makes 16. But 4, the second figure of the divisor, is not contained 7 times in 16, therefore 6 will be the first figure of the quotient.

It is easy to see that this must be 6000, when the division is completed ; because there being five figures in the divisor, and the first figure of the divisor being larger than the first figure of the dividend, we are obliged to take the six first figures of the dividend for the first partial dividend ; and the dividend containing nine figures, the right hand figure of this partial dividend, is in the thousands' place. I write 6 in the quotient, and multiply the divisor by it, and write the result under the dividend, so that the first figure on the right hand may stand under the sixth figure of the dividend, counted from the left, or under the place of thousands. This product, subtracted from the dividend as it stands, leaves a remainder 51518 ; by the side of this I bring down the next figure of the dividend, which is 0, and the second partial dividend is 515180. Trying as before with the 6, and then with the 4, into the first figures of this partial dividend, I find the divisor is contained in it 8 (800) times. I write 8 in the quotient, then multiplying and subtracting as before, I find a remainder 1284. I bring down the next figure of the dividend, which gives 12847 for the next partial dividend. I find that the divisor is not contained in this at all. I put 0 in the quotient, so that the other figures may stand in their proper places, when the division is completed. Then I bring down the next figure of the dividend, which gives for a partial dividend, 128474. The divisor is contained twice in this. Multiplying and subtracting as before, I find no remainder. The division therefore is completed.

Proof. It was observed in the commencement of this Art. that division is proved by multiplying the divisor by the quotient. This is always done during the operation. In the last example, the divisor was first multiplied by 6 (6000,) and then by 8 (800,) and then by 2 ; we have only to add these numbers together in the order they stand in, and if the work is right, this sum will be the dividend. The asterisms show the numbers to be added.

From the above examples we derive the following general rule for division : *Place the divisor at the right of the dividend, separate them by a mark, and draw a line under the divisor, to separate it from the quotient. Take as many figures on the left of the dividend as are necessary to contain the divisor once or more. Seek how many times the first figure of the divisor is contained in the first, or two first figures of these, then increasing the first figure of the divisor by one, seek how many times that is contained in the same figure or figures. Take the figure contained within these limits, which appears the most probable, and multiply the two left hand figures of the divisor by it ; if that is not sufficient to determine, multiply the third, and so on. When the first figure of the quotient is discovered, multiply the divisor by it, and subtract the product from the partial dividend. Then write the next figure of the dividend by the side of the remainder. This is the next partial dividend. Seek as before how many times the divisor is contained in this, and place the result in the quotient, at the right of the other quotient figure, then multiply and subtract, as before ; and so on, until all the figures of the dividend have been used. If it happens that any partial dividend is not so large as the divisor, a zero must be put in the quotient, and the next figure of the dividend written at the right of the partial dividend.*

Note. If the remainder at any time should exceed the

divisor, the quotient figure must be increased, and the multiplication and subtraction must be performed again. If the product of the divisor, by any quotient figure, should be larger than the partial dividend, the quotient figure must be diminished.

Short Division.

When the divisor is a small number, the operation of division may be much abridged, by performing the multiplication and subtraction in the mind without writing the results. In this case it is usual to write the quotient under the dividend. This method is called *short division.*

A man purchased a quantity of flour for 3045 dollars, at 7 dollars a barrel. How many barrels were there ?

Long Division.	Short Division.
3045 (7	3045 (7
28 ———	———
—— 435	435
24	
21	
——	
35	
35	
· ·	

In short division, I say 7 into 30, 4 times; I write 4 underneath; then I say 4 times 7 are 28, which taken from 30 leaves 2. I suppose the 2 written at the left of 4, which makes 24; then 7 into 24, 3 times, writing 3 underneath, I say 3 times 7 are 21, which taken from 24 leaves 3. I suppose the 3 written at the left of 5, which makes 35; then 7 in 35, 5 times exactly; I write 5 underneath, and the division is completed.

If the work in the short and long be compared together, they will be found to be exactly alike, except in the short it is not written down.

X. *How many yards of cloth, at 6 dollars a yard, may be bought for 45 dollars ?*

42 dollars will buy 7 yards, and 48 dollars will buy 8 yards. 45 dollars then will buy more than 7 yards and less than 8 yards, that is, 7 yards and a part of another yard. As cases like this may frequently occur, it is necessary to know what this part is, and how to distinguish one part from another.

When any thing, or any number is divided into two equal parts, one of the parts is called the *half* of the thing or number. When the thing or number is divided into three equal parts, one of the parts is called one *third* of the thing or number ; when it is divided into four equal parts, the parts are called *fourths ;* when into five equal parts, *fifths,* &c. That is the parts always take their names from the number of parts into which the thing or number is divided. It is evident that whatever be the number of parts into which the thing or number is divided, it will take all the parts to make the whole thing or number.' That is, it will take two halves, three thirds, four fourths, five fifths, &c. to make a whole one. It is also evident, that the more parts a thing or number is divided into, the smaller the parts will be. That is, halves are larger than thirds, thirds are larger than fourths, and fourths are larger than fifths, &c.

When a thing or number is divided into parts, any number of the parts may be used. When a thing is divided into three parts, we may use one of the parts or two of them. When it is divided into four parts, we may use one, two, or three of them, and so on. Indeed it is plain, that, when any thing is divided into parts, each part becomes a new unit, and that we may number these parts as well as the things themselves before they were divided.

* Hence we say one third, two thirds, one fourth, two fourths, three fourths, one fifth, two fifths, three fifths, &c.

These parts of one are called *fractions,* or *broken numbers.* They may be expressed by figures as well as whole numbers; but it requires two numbers to express them, one to show into how many parts the thing or number is to be divided (that is, how large the parts are, and how many it takes to make the whole one) ; and the other, to show how many of these parts are used. It is evident that these numbers must always be written in such a manner, that we may know what each of them is intended to represent. It is agreed to write the numbers one above the other, with a line between them. The number below the line shows into how many parts the thing or number is divided, and the number above the line shows how many of the parts are used. Thus ⅔ of an orange signifies, that the orange is divided into three equal parts, and that two of the parts or pieces are used. ⅗ of a yard of cloth, signifies that the yard is supposed to be divided into five equal parts, and that three of these parts are used. The number below the line is called the *denominator*, because it gives the denomination or name to the fraction, as halves, thirds, fourths, &c. and the number above the line is called the *numerator*, because it shows how many parts are used.

We have applied this division to a single thing, but it often happens that we have a number of things which we consider as a bunch or collection, and of which we wish to take parts, as we do of a single thing. In fact it frequently happens that one case gives rise to the other, so that both kinds of division happen in the same question.

If a barrel of cider cost 2 dollars, what will ½ of a barrel cost?

To answer this question, it is evident the number two must be divided into two equal parts, which is very easily done. ½ of 2 is 1.

Again, it may be asked, if a barrel of cider cost 2 dollars, what part of a barrel will one dollar buy?

This question is the reverse of the other. But we have just seen that 1 is ½ of 2, and this enables us to answer the question. It will buy ½ of a barrel.

If a yard of cloth cost 3 dollars, what will ⅓ of a yard cost? What will ⅔ of a yard cost?

If 3 dollars be divided into 3 equal parts, one of the parts will be 1, and two of the parts will be 2. Hence ⅓ of a yard will cost 1 dollar, and ⅔ will cost 2 dollars.

If this question be reversed, and it be asked, what part of a yard can be bought for 1 dollar, and what part for 2 dollars; the answer will evidently be ⅓ of a yard for 1 dollar, and ⅔ for 2 dollars.

It is easy to see that any number may be divided into as many parts as it contains units, and that the number of units used will be so many of the parts of that number. Hence if it be asked, what part of 5, 3 is, we say, $\frac{1}{5}$ of 5, because 1 is $\frac{1}{5}$ of 5, and 3 is three times as much.

We can now answer the question proposed above, viz. How many yards of cloth, at 6 dollars a yard, may be bought for 45 dollars?

42 dollars will buy 7 yards, and the other 3 dollars will buy $\frac{1}{2}$ of a yard.   Ans. $7\frac{1}{2}$ yards, which is read 7 yards and $\frac{1}{2}$ of a yard.

A man hired a labourer for 15 dollars a month; at the end of the time agreed upon, he paid him 143 dollars. How many months did he work?

Operation.

<pre>
                    143 (15
Price of 9 months   135 ——
                    ——   9¹⁸₁₇ months.
        Remainder   8
</pre>

The wages of 9 months is 135 dollars, which subtracted from 143, leaves 8 dollars.   Now 1 dollar will pay for $\frac{1}{15}$ of a month, consequently 8 dollars will pay for $\frac{8}{15}$ of a month. Ans. $9\frac{8}{15}$ months.

Note.   The number which remains after division, as 8 in this example, is called the *remainder.*

At 97 dollars a ton, how many tons of iron may be bought for 2467 dollars?

Operation.

<pre>
      2467 (97
      194   ——
      ——   25⁴²₉₇ tons.
      527
      485
       ——
Remainder   42 dollars.
</pre>

After paying for 25 tons, there are 42 dollars left.   1 dollar will buy $\frac{1}{97}$ of a ton, and 42 dollars will buy $\frac{42}{97}$ of a ton.

How many times is 324 contained in 18364?

Operation.

<pre>
    18364 (324
    1620   ———
    ———   56²²₃₂₄ times.
    2164
    1944
    ———
Remainder   220
</pre>

It is contained 56 times and 220 over. 1 is $\frac{1}{324}$ of 324, and 220 is $\frac{220}{324}$ of '324. Ans. 56 times and $\frac{220}{324}$ of another time.

From the above examples, we deduce the following general rule for the remainder : *When the division is performed, as far as it can be, if there is a remainder, in order to have the true quotient, write the remainder over the divisor in the form of a fraction, and annex it to the quotient.*

XI. We observed in Art. V. that when the multiplier is 10, 100, 1000, &c. the multiplication is performed by annexing the zeros at the right of the multiplicand. In like manner when the divisor is 10, 100, 1000, &c. division may be performed by cutting off as many places from the right of the dividend as there are zeros in the divisor.

At 10 *cents a pound, how many pounds of meat may be bought for* 64 *cents ?*

The 6 which stands in tens' place shows how many times ten is contained in 60, for 60 signifies 6 tens, and the 4 shows how many the number is more than 6 tens, therefore 4 is the remainder. The operation then may be performed thus, 6.4. The answer is $6\frac{4}{10}$ pounds.

A man has 2347 *lb. of tobacco, which he wishes to put into boxes containing* 100 *lb. each ; how many boxes will it take ?*

It is evident that 100 is contained in 2300, 23 times, consequently it will take 23 boxes, and there will be 47 lbs. left, which will fill $\frac{47}{100}$ of another box. The operation may be performed thus, 23.47. Answer $23\frac{47}{100}$.

In general if one figure be cut off from the right, the tens will be brought into the units' place, and hundreds into the tens' place, &c. If two figures be cut off, hundreds are brought into the units' place, and thousands into the tens' place, &c. And if three figures be cut off, thousands are brought into the units' place, &c. that is, the numbers will be made 10, 100, or 1000 times less than before.

Hence to divide by 10, 100, 1000, *&c. cut off from the right of the dividend as many figures as there are zeros in the divisor. The remaining figures will be the quotient, and the figures cut off will be the remainder, which must be written over the divisor, and annexed to the quotient.*

XII.　We observed in Art. X, that any two numbers being given, it is easy to tell what part of the one the other is. Thus :

What part of 10 *yards are* 3 *yards ?　Ans.* 1 *is* $\frac{1}{10}$ *of* 10, *and* 3 *is* $\frac{3}{10}$ *of ten.*

What part of 237 *barrels is* 82 *barrels ?　Ans.* 1 *is* $\frac{1}{237}$ *of* 237, *and* 82 *is* $\frac{82}{237}$ *of* 237.

Fractions are properly parts of a unit, but by extension the term *fraction* is often applied to numbers larger than unity. This happens when the numerator is larger than the denominator, in which case there are more parts taken than are sufficient to make a unit. All fractions in which the numerator is equal to the denominator, as $\frac{2}{2}$, $\frac{3}{3}$, $\frac{4}{4}$, $\frac{11}{11}$, &c. are equal to unity ; all in which the numerator is less than the denominator are less than unity, and are called *proper* fractions ; all in which the numerator is greater than the denominator, are more than unity, and are called *improper* fractions. Thus $\frac{7}{5}$, $\frac{11}{7}$, $\frac{25}{3}$, are improper fractions.

The process of finding what part of one number another number is, is called finding their *ratio.*

What is the ratio of 5 bushels to 3 bushels, or of 5 to 3 ? This is the same as to say, what part of 5 is 3 ? The answer is $\frac{3}{5}$. The ratio of 5 to 3 is $\frac{3}{5}$.

What part of 3 *is* 5 ?　*Answer* $\frac{5}{3}$.　*The ratio of* 3 *to* 5 *is* $\frac{5}{3}$.

What is the ratio of 35 *yards to* 17 *yards.　Answer* $\frac{17}{35}$.

What is the ratio of 17 *to* 35 ?　*Answer* $\frac{35}{17}$.

To find what part of one number another is, make the number which is called the part (whether it be the larger or smaller) the numerator of a fraction, and the other number the denominator.

Also to find the ratio of one number to another, make the number which is expressed first the denominator, and the other the numerator.

XIII.　*A gentleman gave* $\frac{1}{5}$ *of a dollar each to* 17 *poor persons ; how many dollars did it take ?*

It took $\frac{17}{5}$ of a dollar. But $\frac{5}{5}$ of a dollar make a dollar, consequently as many times as 5 is contained in 17, so many dollars it is. 5 is contained 3 times in 17, and 2 over.

14

That is, $\frac{17}{5}$ make 3 dollars, and there are $\frac{2}{5}$ of another dol-lar. Ans. $3\frac{2}{5}$ dollars.   -

If 1 man consume $\frac{1}{35}$ of a barrel of flour in a week, how many barrels will an army of 537 men consume in the same time ?

They will consume $\frac{537}{35}$. $\frac{35}{35}$ of a barrel make a barrel, therefore as many times as 35 is contained in 537, so many barrels it is.

```
537  (35
35    ——
——   15 1 2  barrels.   Ans.
187      3 5
175

——
12
```

35 is contained 15 times in 537 and 12 over, which is $\frac{12}{35}$ of another barrel.

Numbers like $3\frac{2}{5}$, $15\frac{12}{35}$, which contain a whole number and a fraction, are called *mixed* numbers. The above pro-cess by which $\frac{17}{5}$ was changed to $3\frac{2}{5}$, and $\frac{537}{35}$ to $15\frac{12}{35}$, is called reducing *improper* fractions to *whole* or *mixed* num-bers.

Since the denominator always shows how many of the parts make a whole one, it is evident that any *improper* frac-tion may be reduced to a whole or mixed number, by the fol-lowing rule : *Divide the numerator by the denominator, and the quotient will be the whole number. If there be a remain-der, write it over the denominator, and annex it to the quo-tient, and it will form the mixed number required.*

XIV. It is sometimes necessary to change a whole or a mixed number to an improper fraction.

A man distributed 3 dollars among some beggars, giving them $\frac{1}{5}$ of a dollar apiece ; how many received the money ? That is, in 3 dollars, how many fifths of a dollar ?

Each dollar was divided equally among 5 persons, conse-quently 3 dollars were given to 15 persons. That is, 3 dol-lars are equal to $\frac{15}{5}$ of a dollar.

A man distributed $18\frac{3}{7}$ bushels of wheat among some poor persons, giving them $\frac{1}{7}$ of a bushel each ; how many persons were there ?

This question is the same as the following:

In $18\frac{3}{7}$ *bushels, how many* $\frac{1}{7}$ *of a bushel? That is, how many 7ths of a bushel?*

In 1 bushel there are $\frac{7}{7}$, consequently in 18 bushels there are 18 times 7 sevenths; that is, $\frac{126}{7}$, and $\frac{3}{7}$ more make $\frac{129}{7}$. Answer 129 persons.

Reduce $28\frac{17}{25}$ *to an improper fraction. That is, in* $28\frac{17}{25}$ *how many* $\frac{1}{25}$.

Since there are $\frac{25}{25}$ in 1, in 28 there must be 28 times as many. 28 times 25 are 700, and 17 more are 717. Ans. $\frac{717}{25}$.

Hence to reduce a whole number to an improper fraction with a given denominator, or a mixed number to an improper fraction: *multiply the whole number by the denominator, and if it is a mixed number add the numerator of the fraction, and write the result over the denominator.*

XV. *A man hired 7 labourers for 1 day, and gave them* $\frac{3}{5}$ *of a dollar apiece; how many dollars did he pay the whole?*

If we suppose each dollar to be divided into 5 equal parts, it would take 3 parts to pay 1 man, 6 parts to pay 2 men, &c. and 7 times 3 or 21 parts, that is, $\frac{21}{5}$ of a dollar to pay the whole. $\frac{21}{5}$ of a dollar are $4\frac{1}{5}$ dollars. Ans. $4\frac{1}{5}$ dollars.

A man bought 13 bushels of grain, at $\frac{5}{6}$ *of a dollar a bushel; how many dollars did it come to?*

$\frac{5}{6}$ of a dollar are 5 shillings. 13 bushels at 5 shillings a bushel, would come to 65 shillings, which is 10 dollars and 5 shillings.

In the first form, 13 times $\frac{5}{6}$ of a dollar are $\frac{65}{6}$ of a dollar; that is $10\frac{5}{6}$ dollars, as before.

A man found by experience, that one day with another, his horse would consume $\frac{13}{37}$ *of a bushel of oats in a day; how many bushels would he consume in 5 weeks or 35 days?*

If we suppose each bushel to be divided into 37 equal parts, he would consume 13 parts each day. In 35 days he would consume 35 times 13 parts, which is 455 parts. But the parts are 37ths, therefore it is $\frac{455}{37} = 12\frac{11}{37}$ bushels.

$$35$$
$$13$$
$$\overline{}$$
$$105$$
$$35$$
$$\overline{}$$
$$455$$

$$\tfrac{455}{37} = 12\tfrac{11}{37}$$

This process is called *multiplying a fraction by a whole number.*

Multiply $\tfrac{253}{1372}$ *by* 48.

The fraction $\tfrac{253}{1372}$ signifies that 1 is divided into 1372 equal parts, and that 253 of those parts are used. To multiply it by 48, is to take 48 times as many parts, that is, to multiply the numerator 253 by 48·

$$253$$
$$48$$
$$\overline{}$$
$$2024$$
$$1012$$
$$\overline{}$$
$$12144$$

$$\tfrac{12144}{1372} = 8\tfrac{1168}{1372}$$

The product of 253 by 48 is 12144; this written over the denominator is $\tfrac{12144}{1372}$, which being reduced is $8\tfrac{1168}{1372}$. Ans.

To multiply a fraction then, is to multiply the number of parts used; hence the rule: *multiply the numerator and write the product over the denominator.*

Note. It is generally most convenient, when the numerator becomes larger than the denominator, to reduce the fraction to a whole or mixed number.

It is sometimes necessary to multiply a mixed number.

Bought 13 tons of iron, at $97\tfrac{1}{37}$ *dollars a ton; what did it come to?*

In this example the whole number and the fraction must be multiplied separately. 13 times 97 are 1261. 13 times $\tfrac{1}{37}$ are $\tfrac{13}{37}$, equal to $4\tfrac{3}{37}$; this added to 1261 makes $1265\tfrac{3}{37}$ dollars. Ans.

Operation.

```
97        14
13        13
──        ──
291       42
 97       14
────      ───
1261      182
```

$\frac{182}{13} = 4\frac{10}{13}$

$1261 + 4\frac{10}{13} = 1265\frac{10}{13}$ dolls.

Hence, to multiply a mixed number : *multiply the whole number and the fraction separately ; then reduce the fraction to a whole or mixed number, and add it to the product of the whole number.*

.XVI. We have seen that single things may be divided into parts, and that numbers may be divided into as many parts as they contain units ; that is, 4 may be divided into 4 parts, 7 into 7 parts, &c. It now remains to be shown, how every number may be divided into any number of equal parts.

If 3 yards of cloth cost 12 dollars, what is that a yard ?

It is evident that the price of 3 yards must be divided into 3 equal parts, in order to have the price of 1 yard. That is, ⅓ of 12 must be found.

We know by the table of Pythagoras, that 3 times 4 are 12, therefore ⅓ of 12, or 4 dollars is the price of 1 yard.

If 5 yards of cloth cost 45 dollars, what is that a yard ?

1 yard will cost ⅕ of 45 dollars. 5 times 9 are 45, therefore 9 is ⅕ of 45, or the price of 1 yard.

The two last examples are similar to the first example Art. 9. If we take 1 dollar for each yard, it will be 5 dollars, then one for each yard again, it will be 5 more, and so on, until the whole is divided. The question, therefore, is to see how many times 5 is contained in 45, and the result will be a number that is contained 5 times in 45. 5 is contained 9 times, therefore 9 is contained 5 times in 45. This is evident also from Art. III. *When a number, therefore, is to be divided into parts, it is done by division. The number to be divided is the dividend, the number of parts the divisor, and the quotient is one of the parts.*

14 *

A man owned a share in a bank worth 136 *dollars, and sold* ½ *of it; how many dollars did he sell it for?*

136 (2

Ans. 68 dollars.

2 is contained 68 times in 136, therefore 2 times 68 are 136, consequently 68 is ½ of 136.

A ticket drew a prize of 2,845 *dollars, of which A owned* ⅕; *what was his share?*

2845 (5

Ans. 569 dollars.

Since 5 is contained 569 times in 2,845, 5 times 569 are equal to 2,845, therefore 569 is ⅕ of 2,845. Division may be explained, as taking a part of a number. In the above example I say, ⅕ of 28(00) is 5(00) and 3(00) over. Then supposing 3 at the left of 4, I say, ⅕ of 34(0) is 6(0) and 4(0) over. Then ⅕ of 45 is 9. Writing all together it makes 569, as before. The same explanation will apply when the divisor is a large number.

Bought 43 *tons of iron for* 4,171 *dollars; how much was it a ton?*

1 ton is $\frac{1}{43}$ part of 43 tons, therefore the price of 1 ton will be $\frac{1}{43}$ part of the price of 43 tons.

```
4171 (43
387
----  97 dollars.
 301
 301
 ---
```

Two men A and B traded in company and gained 456 *dollars, of which A was to have* ⅝ *and B* ⅜; *what was the share of each?*

The name of the fraction shows how to perform this example. ⅝ of 456 signifies that 456 must be divided into 8 equal parts, and 5 of the parts taken. ⅛ of 456 is 57, 5 times 57 are 285, and 3 times 57 are 171. A's share 285, and B's 171 dollars.

$$456\ (8 \qquad\qquad 57$$
$$\qquad\qquad\qquad\qquad\qquad\quad 3$$
$$57$$
$$5 \qquad\qquad\qquad \text{B's share } 171 \text{ dollars.}$$

A's share 285 dollars.

A man bought 68 barrels of pork for 1224 dollars, and sold 47 barrels, at the same rate that he gave for it. How much did the 47 barrels come to ?

To answer this question it is necessary to find the price of 1 barrel, and then of 47. 1 barrel costs $\frac{1}{68}$ of 1224 dollars, and 47 barrels cost $\frac{47}{68}$ of it. $\frac{1}{68}$ of 1224 is 18. 47 times are 18 are 846. Ans. 846 dollars.

To find any fractional part of a number, divide the number by the denominator of the fraction, and multiply the quotient by the numerator.

A man bought 5 yards of cloth for 28 dollars ; what was that a yard ?

$\frac{1}{5}$ of 25 is 5, and $\frac{1}{5}$ of 30 is 6. $\frac{1}{5}$ of 28 then must be between 5 and 6.

Cases of this kind frequently occur, in which a number cannot be divided into exactly the number of parts proposed, except by taking fractions. But it may easily be done by fractions.

$\frac{1}{5}$ of 25 dollars is 5 dollars. It now remains to find $\frac{1}{5}$ of 3 dollars. Suppose each dollar divided into 5 equal parts, and take 1 part from each. That will be 3 parts or $\frac{3}{5}$ of a dollar. Ans. $5\frac{3}{5}$ dollars. $\frac{3}{5}$ of a dollar is $\frac{3}{5}$ of 100 cents, which is 60 cents. Ans. $5.60.

A man had 853 lb. of butter, which he wished to divide into 7 equal parts ; how many pounds would there be in each part ?

$\frac{1}{7}$ of 847 lb. is 121 lb. Then suppose each of the 6 remaining pounds to be divided into 7 equal parts, and take 1 part from each ; that will be 6 parts, or $\frac{6}{7}$ of a pound. Ans. $121\frac{6}{7}$.

$$853\ (7$$
$$\overline{\qquad\qquad}$$
$$121\frac{6}{7} \text{ lb. \quad Ans.}$$

*A man having travelled 47 days, found that he had travel-
led 1800 miles; how many miles had he travelled in a day
on an average? How many miles would . he travel in 53
days, at that rate?*

In one day he travelled $\frac{1}{47}$ of 1800 miles, and in 53 days
he would travel $\frac{53}{47}$ of it. $\frac{1}{47}$ of 1800 is 38, and 14 over.
$\frac{1}{47}$ of 1 is $\frac{1}{47}$, $\frac{1}{47}$ of 14 is 14 times as much, that is, $\frac{14}{47}$. In
one day he travelled $38\frac{14}{47}$ miles. In 53 days he would
travel 53 times $38\frac{14}{47}$ miles.

```
1800 (47              38         53
141                   53         14
 ──── 38¹⁴₄₇ miles in 1 day. ──   ──
  390                114         212
  376                190         53
  ────               ────        ────
   14                2014        742
              +       15³⁷₄₇           ⁷⁴²₄₇ = 15³⁷₄₇
                     ─────
         Ans.   2029³⁷₄₇ miles in 53 days.
```

Hence to divide a number into parts; *divide it by the
number of parts required, and if there be a remainder, make
it the numerator of a fraction, of which the divisor is the denominator.*

N. B. This rule is substantially the same as the rule in
Art. X.

When one part is found, any number of the parts may be
found by multiplication.

It was shown in Art. X. that, in a fraction, the denominator shows into how many parts 1 is supposed to be divided,
and that the numerator shows how many of the parts are used.
It will appear from the following examples, that the numerator
is a dividend, and the denominator a divisor, and that the
fraction expresses a quotient. The denominator shows into
how many parts the numerator is to be divided. In this manner division may be expressed without being actually performed. If the fraction be multiplied or divided, the quotient will also be multiplied or divided. Hence division may
be first expressed, and the necessary operations performed on
the quotient, and the operation of division itself omitted,
until the last, which is often more convenient. Also, when
the divisor is larger than the dividend, division may be expressed, though it cannot be performed.

A gentleman wishes to divide 23 *barrels of flour equally among* 57 *families ; how much must he give them apiece ?*

In this example, the divisor 57 is greater than the dividend 23. If he had only 1 barrel to divide, he could give them only $\frac{1}{57}$ of a barrel apiece ; but since he had 23 barrels, he can give each 23 times as much, that is, $\frac{23}{57}$ of a barrel.

Hence it appears that $\frac{23}{57}$ rightly expresses the quotient of 23 by 57.

If it be asked how many times is 57 contained in 23 ? It is not contained one time, but $\frac{23}{57}$ of one time.

If 10 *lbs. of copper cost* 3 *dollars, what is it per lb. ?*

Here 3 must be divided by 10. $\frac{1}{10}$ of 1 is $\frac{1}{10}$, and $\frac{1}{10}$ of 3 must be $\frac{3}{10}$. Ans. $\frac{3}{10}$ of a dollar, that is, 30 cents.

At 43 *dollars per hhd., what would be the price of* 25 *galls. of gin ?*

25 galls. are $\frac{25}{63}$ of a hogshead. To find the price of 1 gallon is to find $\frac{1}{63}$ of 43 dolls., and to find the price of 25 galls. is to find $\frac{25}{63}$ of 43 dolls. $\frac{1}{63}$ of 1 is $\frac{1}{63}$, $\frac{1}{63}$ of 43 is 43 times as much, that is, $\frac{43}{63}$. $\frac{25}{63}$ is 25 times as much as $\frac{1}{63}$, that is, 25 times $\frac{43}{63}$. 25 times $\frac{43}{63}$ are $\frac{1075}{63} = 17\frac{4}{63}$ dolls. Ans.

If 5 *tons of hay cost* 138 *dolls. what cost* 3 *tons ?*

3 tons will cost $\frac{3}{5}$ of 138 dolls. This may be done as follows. $\frac{1}{5}$ of 138 is $27\frac{3}{5}$, and 3 times $27\frac{3}{5}$, are $82\frac{4}{5}$ dolls. Ans. Or,

Expressing the division, instead of performing it, $\frac{1}{5}$ of 138 is $\frac{138}{5}$. $\frac{3}{5}$ of 138 are 3 times $\frac{138}{5}$, that is, $\frac{414}{5} = 82\frac{4}{5}$ dolls. as before.

Note. $\frac{1}{5}$ of 138 by the above rule is $27\frac{3}{5}$. But the same result will be obtained, if we say, $\frac{1}{5}$ of 138 is $\frac{138}{5}$, for $\frac{138}{5}$ are equal to $27\frac{3}{5}$.

The process in this Art. is called *multiplying a whole number by a fraction.* Multiplication strictly speaking is repeating the number a certain number of times, but by extension, it is made to apply to this operation. The definition of multiplication, in its most extensive sense, is *to take one number, as many times as one is contained in another number.* Therefore if the multiplier be greater than 1, the product will be greater than the multiplicand ; but if the multiplier be only a part of 1, the product will be only a part of the multiplicand.

It was observed in Art. III. that when two whole numbers are to be multiplied together. either of them may be made the multiplier, without affecting the result. In the same manner, to multiply a whole number by a fraction, is the same as to multiply a fraction by a whole number.

For in the last example but one, in which 43 was multiplied by $\frac{25}{63}$, 25 and 43 were multiplied together, and the product written over the denominator 63, thus $\frac{1075}{63}$. The same would have been done, if $\frac{25}{63}$ had been multiplied by 43.

In the last example also, 138 was multiplied by $\frac{3}{5}$. The result would have been the same if $\frac{3}{5}$ had been multiplied by 138.

This may be proved directly.

It is required to find $\frac{25}{63}$ of 43. $\frac{25}{63}$ of 1 is $\frac{25}{63}$, $\frac{25}{63}$ of 43 must be 43 times as much, that is, 43 times $\frac{25}{63}$, or $\frac{1075}{63} = 17\frac{4}{63}$. So also $\frac{3}{5}$ of 1 is $\frac{3}{5}$. $\frac{3}{5}$ of 138 must be 138 times as much, that is, 138 times $\frac{3}{5}$, or $\frac{414}{5} = 82\frac{4}{5}$.

Hence to multiply a fraction by a whole number, or a whole number by a fraction; multiply the whole number and the numerator of the fraction together, and write the product over the denominator of the fraction.

XVII. *If 3 yards of cloth cost $\frac{3}{4}$ of a dollar, what is that a yard?*

$\frac{3}{4}$ are 3 parts. $\frac{1}{3}$ of 3 parts is 1 part. Ans. $\frac{1}{4}$ of a dollar.

A man divided $1\frac{5}{7}$ of a barrel of flour equally among 4 families; how much did he give them apiece?

$1\frac{5}{7}$ are 12 parts. $\frac{1}{4}$ of 12 parts is 3 parts. Ans. $\frac{3}{7}$ of a barrel each.

This process is dividing a fraction by a whole number. A fraction is a certain number of parts. It is evident that any number of these parts may be divided into parcels, as well as the same number of whole ones. The numerator shows how many parts are used; therefore *to divide a fraction, divide the numerator.*

But it generally happens that the numerator cannot be exactly divided by the number, as in the folllowing example.

A boy wishes to divide $\frac{3}{4}$ of an orange equally between two other boys; how much must he give them apiece?

If he had 3 oranges to divide, he might give them 1 apiece, and then divide the other into two equal parts, and give one part to each, and each would have $1\frac{1}{2}$ orange. Or he might cut them all into two equal parts each, which would make six parts, and give 3 parts to each, that is, $\frac{3}{2} = 1\frac{1}{2}$, as before. But according to the question, he has $\frac{3}{4}$ or 3 pieces, consequently he may give 1 piece to each, and then cut the other into two equal parts, and give 1 part to each, then each will have $\frac{1}{4}$ and $\frac{1}{2}$ of $\frac{1}{4}$. But if a thing be cut into four equal parts, and then each part into two equal parts, the whole will be cut into 8 equal parts or eighths; consequently $\frac{1}{2}$ of $\frac{1}{4}$ is $\frac{1}{8}$. Each will have $\frac{1}{4}$ and $\frac{1}{8}$ of an orange. Or he may cut each of the three parts into two equal parts, and give $\frac{1}{2}$ of each part to each boy, then each will have 3 parts, that is $\frac{3}{8}$. Therefore $\frac{1}{2}$ of $\frac{3}{4}$ is $\frac{3}{8}$. Ans. $\frac{3}{8}$.

A man divided $\frac{1}{5}$ of a barrel of flour equally between 2 labourers; what part of the whole barrel did he give to each?

To answer this question it is necessary to find $\frac{1}{2}$ of $\frac{1}{5}$. If the whole barrel be divided first into 5 equal parts or fifths, and then each of these parts into 2 equal parts, the whole will be divided into 10 equal parts. Therefore, $\frac{1}{2}$ of $\frac{1}{5}$ is $\frac{1}{10}$. He gave them $\frac{1}{10}$ of a barrel apiece.

A man owning $\frac{7}{8}$ of a share in a bank, sold $\frac{1}{3}$ of his part; what part of the whole share did he sell?

If a share be first divided into 8 equal parts, and then each part into 3 equal parts, the whole share will be divided into 24 equal parts. Therefore $\frac{1}{3}$ of $\frac{1}{8}$ is $\frac{1}{24}$, and $\frac{1}{3}$ of $\frac{7}{8}$ is 7 times as much, that is, $\frac{7}{24}$. Ans. $\frac{7}{24}$.

Or since $\frac{1}{8} = \frac{3}{24}$, $\frac{7}{8} = \frac{21}{24}$, and $\frac{1}{3}$ of $\frac{21}{24} = \frac{7}{24}$.

In the three last examples the division is performed by multiplying the denominator. In general, if the denominator of a fraction be multiplied by 2, the unit will be divided into twice as many parts, consequently the parts will be only one half as large as before, and if the same number of the small parts be taken, as was taken of the large, the value of the fraction will be one half as much. If the denominator be multiplied by three, each part will be divided into three parts, and the same number of the parts being taken, the fraction will be one third of the value of the first. Finally, if the denominator be multiplied by any number, the parts will be so many times smaller. Therefore, *to divide a frac-*

tion, if the numerator cannot be divided exactly by the divisor, multiply the denominator by the divisor.

A man divided $\frac{5}{8}$ of a hogshead of wine into 7 equal parts, in order to put it into 7 vessels; what part of the whole hogshead did each vessel contain?

The answer, according to the above rule, is $\frac{5}{112}$. The propriety of the answer may be seen in this manner. Suppose each 16th to be divided into 7 equal parts, the parts will be 112ths. From each of the $\frac{5}{16}$ take one of the parts, and you will have 5 parts, that is $\frac{5}{112}$.

A man owned $\frac{7}{18}$ of a ship's cargo; but in a gale the captain was obliged to throw overboard goods to the amount of $\frac{4}{9}$ of the whole cargo. What part of the loss must this man sustain?

It is evident that he must lose $\frac{4}{9}$ of his share, that is, $\frac{4}{9}$ of $\frac{7}{18}$.

$\frac{1}{9}$ of $\frac{1}{18} = \frac{1}{162}$, $\frac{1}{9}$ of $\frac{7}{18} = \frac{7}{162}$, and $\frac{4}{9}$ must be 4 times as much, that is, $\frac{28}{162}$. Ans. $\frac{28}{162}$ of the whole loss.

Or it may be said, that since he owned $\frac{7}{18}$ of the ship, he must sustain $\frac{7}{18}$ of the loss, that is, $\frac{7}{18}$ of $\frac{4}{9}$. $\frac{1}{18}$ of $\frac{1}{9} = \frac{1}{162}$, $\frac{1}{18}$ of $\frac{4}{9} = \frac{4}{162}$, and $\frac{7}{18}$ is 7 times as much, that is, $\frac{28}{162}$, as before.

This process is *multiplying one fraction by another*, and is similar to multiplying a whole number by a fraction, Art. XVI. If the process be examined, it will be found that the denominators were multiplied together for a new denominator, and the numerators for a new numerator. In fact to take a fraction of any number, is to divide the number by the denominator, and to multiply the quotient by the numerator. But a fraction is divided by multiplying its denominator, and multiplied by multiplying its numerator. We have seen in the above example, that when two fractions are to be multiplied, either of them may be made multiplier, without affecting the result. Therefore, to take a fraction of a fraction, *that is, to multiply one fraction by another, multiply the denenominators together for a new denominator, and the numerators for a new numerator.*

If 7 dollars will buy $5\frac{3}{8}$ bushels of rye, how much will 1 dollar buy? How much will 15 dollars buy?

1 dollar will buy $\frac{1}{7}$ of $5\frac{3}{8}$ bushels. In order to find $\frac{1}{7}$ of it, $5\frac{3}{8}$ must be changed to eighths, $5\frac{3}{8} = \frac{43}{8}$. $\frac{1}{7}$ of $\frac{43}{8} = \frac{43}{56}$. 1 dollar will buy $\frac{43}{56}$ of a bushel. 15 dollars will buy 15

times as much. 15 times $\frac{43}{50} = \frac{645}{50} = 11\frac{39}{50}$. Ans. $11\frac{39}{50}$ bushels.

If 13 *bbls. of beef cost* $95\frac{7}{8}$ *dollars, what will* 25 *bbls. cost ?*

1 bbl. will cost $\frac{1}{13}$ of $95\frac{7}{8}$ dollars, and 25 bbls. will cost $\frac{25}{13}$ of it. To find this, it is best to multiply first by 25, and then divide by 13. For $\frac{25}{13}$ of $95\frac{7}{8}$ is the same as $\frac{1}{13}$ of 25 times $95\frac{7}{8}$.

Operation.

$$95\frac{7}{8} \times 25 = 2396\frac{7}{8}. \quad 2396\frac{7}{8} \;(13$$

$$13$$
$$\overline{}$$
$$109 \qquad\qquad 184\frac{39}{104}$$
$$104$$
$$\overline{}$$
$$56$$
$$52$$
$$\overline{}$$

$$4\frac{7}{8} = \frac{39}{8}. \quad \text{Ans. } 184\frac{39}{104} \text{ dolls.}$$

In this example I divide $2396\frac{7}{8}$ by 13. I obtain a quotient 184, and a remainder $4\frac{7}{8}$, which is equal to $\frac{39}{8}$. Then $\frac{39}{8}$ divided by 13, gives $\frac{39}{104}$, which I annex to the quotient, and the division is completed.

The examples hitherto employed to illustrate the division of fractions, have been such as to require the division of the fractions into parts. It has been shown (Art. XVI.) that the division of whole numbers is performed in the same manner, whether it be required to divide the number into parts, or to find how many times one number is contained in another. It will now be shown that the same is true with regard to fractions.

At 3 *dollars a barrel, how many barrels of cider may be bought for* $8\frac{3}{5}$ *dollars ?*

The numbers must be reduced to fifths, for the same reason that they must be reduced to pence, if one of the numbers were given in shillings and pence.

$3 = \frac{15}{5}$, and $8\frac{3}{5} = \frac{43}{5}$. As many times as $\frac{15}{5}$ are contained in $\frac{43}{5}$, that is, as many times as 15 are contained in 43, so many barrels may be bought.

Expressing the division $\frac{43}{15} = 2\frac{13}{15}$. Ans. $2\frac{13}{15}$ barrels. This result agrees with the manner explained above. For $8\frac{3}{5}$ was reduced to fifths, and the denominator 15 was formed by multiplying the denominator 5 by the divisor 3.

15

How many times is 2 contained in $\frac{4}{7}$?

$2 = \frac{14}{7}$; 14 is contained in 5, $\frac{5}{14}$ of one time. The same result may be produced by the other method.

XVIII. We have seen that a fraction may be divided by multiplying its denominator, because the parts are made smaller. On the contrary, a fraction may be multiplied by dividing the denominator, because the parts will be made larger. If the denominator be divided by 2, for instance, the denominator being rendered only half as large, the unit will be divided into only one half as many parts, consequently the parts will be twice as large as before. If the denominator be divided by 3, the unit will be divided into only one third as many parts, consequently the parts will be three times as large as before, and if the same number of these parts be taken, the value of the fraction will be three times as great, and so on.

If 1 lb. of sugar cost $\frac{1}{8}$ of a dollar, what will 4 lb. cost ?

If the denominator 8 be divided by 4, the fraction becomes $\frac{1}{2}$; that is, the dollar, instead of being divided into 8 parts, is divided into only 2 parts. It is evident that halves are 4 times as large as eighths, because if each half be divided into 4 parts, the parts will be eighths. Ans. $\frac{1}{2}$ doll.

If it be done by multiplying the numerator, the answer is $\frac{4}{8}$, which is the same as $\frac{1}{2}$, for $\frac{8}{8} = 1$, and $\frac{1}{2}$ of $\frac{8}{8} = \frac{4}{8}$.

If 1 lb. of figs cost $\frac{3}{28}$ of a dollar, what will 7 lb. cost ?

Dividing the denominator by 7, the fraction becomes $\frac{3}{4}$. Now it is evident that fourths are 7 times as large as twenty-eighths, because if fourths be divided into 7 parts, the parts will be twenty-eighths. Ans. $\frac{3}{4}$ dolls.

Or multiplying the numerator, 7 times $\frac{3}{28}$ is $\frac{21}{28}$. But $\frac{1}{4}$ $= \frac{7}{28}$, and $\frac{3}{4} = \frac{21}{28}$, so that the answers are the same.

Therefore, *to multiply a fraction, divide the denominator, when it can be done without a remainder.*

Two ways have now been found to multiply fractions, and two ways to divide them.

To multiply a fraction	Multiply	The numerator, Art. 15.
To divide a fraction		The denominator, Art. 17.
To divide a fraction	Divide	The numerator, Art. 17.
To multiply a fraction		The denominator, Art. 18

XIX. We observed a remarkable circumstance in the last article, viz. that $\frac{1}{2} = \frac{4}{8}$ and $\frac{3}{4} = \frac{21}{28}$. This will be found very important in what follows.

A man having a cask of wine, sold $\frac{1}{2}$ of it at one time, and $\frac{1}{3}$ of it at another, how much had he left?

$\frac{1}{2}$ and $\frac{1}{3}$ cannot be added together, because the parts are of different values. Their sum must be more than $\frac{2}{3}$, and less than $\frac{2}{2}$ or 1. If we have dollars and crowns to add together, we reduce them both to pence. Let us see if these fractions cannot be reduced both to the same denomination. Now $\frac{1}{2} = \frac{2}{4} = \frac{3}{6} = \frac{4}{8}$, &c. And $\frac{1}{3} = \frac{2}{6} = \frac{3}{9}$, &c. It appears, therefore, that they may both be changed to sixths. $\frac{1}{2} = \frac{3}{6}$ and $\frac{1}{3} = \frac{2}{6}$, which added together make $\frac{5}{6}$. He had sold $\frac{5}{6}$ and had $\frac{1}{6}$ left.

A man sold $\frac{3}{5}$ of a barrel of flour at one time, and $\frac{4}{7}$ at another, how much did he sell in the whole?

Fifths and *sevenths* are different parts, but if a thing be first divided into 5 equal parts, and then those parts each into 7 equal parts, the parts will be *thirty-fifths.* Also if the thing be divided first into 7 equal parts, and then those parts each into 5 equal parts, the parts will be *thirty-fifths.* Therefore, the parts will be alike. But in dividing them thus, $\frac{3}{5}$ will make $\frac{21}{35}$, and $\frac{4}{7}$ will make $\frac{20}{35}$, and the two added together make $\frac{41}{35}$, that is, $1\frac{6}{35}$. Ans. $1\frac{6}{35}$ barrel.

When the denominators of two or more fractions are alike, they are said to have a *common* denominator. And the process by which they are made alike, is called *reducing* them to a *common* denominator.

In order to reduce pounds to shillings, we multiply by 20, and to reduce guineas to shillings, we multiply by 28. In like manner to reduce two or more fractions to a common denominator, it is necessary to find what denomination they may be reduced to, and what number the parts of each must be multiplied by, to reduce them to that denomination.

If the denominator of a fraction be multiplied by 2, it is the same as if each of the parts were divided into 2 equal parts, therefore it will take 2 parts of the latter kind to make 1 of the former. If the denominator be multiplied by 3, it is the same as if the parts were divided each into 3 equal parts, and it will take 3 parts of the latter kind, to make 1 of the former. Indeed, whatever number the denominator be multiplied by, it is the same as if the parts were each divided into so many equal parts, and it will take so many parts of the latter kind to make 1 of the former. Therefore, to find what the parts must be multiplied by, it is necessary to find what the denominator must be multiplied by to produce the denominator required.

The common denominator then, (which must be found first) must be a number of which the denominators of all the fractions to be reduced, are factors. We shall always find such a number, by multiplying the denominators together. Hence if there are only two fractions, the denominators being multiplied together for the common denominator, the parts of one fraction must be multiplied by the denominator of the other. If there be more than two fractions, since by multiplying all the denominators together, the denominator of each will be multiplied by all the others, the parts in each fraction, that is, the numerators must be multiplied by the denominators of the other fractions.

In the above example to reduce $\frac{3}{5}$ and $\frac{4}{7}$ to a common denominator, 7 times 5 are 35 ; 7 is the number by which the first denominator 5 must be multiplied to produce 35, and consequently the number by which the numerator 3 must be multiplied. 5 is the number, by which 7, the second denominator, must be multiplied to produce 35, and consequently the number by which the numerator 4 must be multiplied.

N. B. It appears from the above reasoning, that if both the numerator and denominator of any fraction be multiplied by the same number, the value of the fraction will remain the same. It will follow also from this, that if both numerator and denominator can be divided by the same number, without a remainder, the value of the fraction will not be altered. In fact, if the numerator be divided by any number, as 3 for example, it is taking $\frac{1}{3}$ of the number of parts ; then if the denominator be divided by 3, these parts will be made 3 times as large as before, consequently the value will be the same as at first. This enables us frequently, when a fraction is expressed with large numbers, to reduce it, and express it with much smaller numbers, which often saves a great deal of labour in the operations.

Take for example $\frac{15}{35}$. Dividing the numerator by 5, we take $\frac{1}{5}$ of the parts, then dividing the denominator by 5, the parts are made 5 times as large, and the fraction becomes $\frac{3}{7}$, the same value as $\frac{15}{35}$. This is called *reducing fractions to lower terms.* Hence

To reduce a fraction to lower terms, *divide both the nume-*

rator and denominator by any number that will divide them both without a remainder.

Note. This gives rise to a question, how to find the divisors of numbers. These may frequently be found by trial. The question will be examined hereafter.

A man bought 4 pieces of cloth, the first contained $23\frac{6}{8}$ yards; the second $28\frac{8}{12}$; the third $37\frac{3}{15}$; and the fourth $17\frac{1}{2}$. How many yards in the whole?

The fractional parts of these numbers cannot be added together until they are reduced to a common denominator. But before reducing them to a common denominator, I observe that some of them may be reduced to lower terms, which will render it much easier to find the common denominator. In $\frac{6}{8}$ the numerator and denominator may both be divided by 2, and it becomes $\frac{3}{4}$. $\frac{8}{12}$ may be reduced to $\frac{2}{3}$, and $\frac{3}{15}$ to $\frac{1}{5}$. I find also that halves may be reduced to fourths, therefore I have only to find the common denominator of the three first fractions, and the fourth can be reduced to the same.

Multiplying the denominators together $3 \times 4 \times 5 = 60$. The common denominator is 60. Now 3 is multiplied by 4 and by 5 to make 60, therefore, the numerator of $\frac{2}{3}$ must be multiplied by 4 and by 5, or, which is the same thing, by 20, which makes 40, $\frac{2}{3} = \frac{40}{60}$. In $\frac{3}{4}$, the four is multiplied by 3 and 5 to make 60, therefore these are the numbers by which the numerator 3 must be multiplied. $\frac{3}{4} = \frac{45}{60}$. In the fraction $\frac{1}{5}$, the 5 is multiplied by 3 and 4 to make 60, therefore these are the numbers by which the numerator 1 must be multiplied. $\frac{1}{5} = \frac{12}{60}$. $\frac{1}{2} = \frac{30}{60}$. These results may be verified, by taking $\frac{2}{3}$, $\frac{3}{4}$, and $\frac{1}{5}$ of 60. It will be seen that $\frac{1}{3}$ of 60 is 20, the product of 4 and 5; $\frac{1}{4}$ of 60 is 15, the product of 3 and 5; and $\frac{1}{5}$ of 60 is 12, the product of 3 and 4. Now the numbers may be added as follows:

$$
\begin{array}{llll}
23\frac{6}{8} & = 23\frac{3}{4} & = 23\frac{45}{60} & 45 \\
28\frac{8}{12} & = 28\frac{2}{3} & = 28\frac{40}{60} & 40 \\
37\frac{3}{15} & = 37\frac{1}{5} & = 37\frac{12}{60} & 12 \\
17\frac{1}{2} & = & 17\frac{30}{60} & 30 \\
\hline
\end{array}
$$

Ans. $107\frac{7}{60}$ yards. 127 $\frac{127}{60} = 2\frac{7}{60}$.

I add together the fractions, which make $\frac{127}{60} = 2\frac{7}{60}$. I write the fraction $\frac{7}{60}$, and add the 2 whole ones with the others.

A man having 23⅔ *barrels of flour, sold* 8⅝ *barrels ; how many barrels had he left ?*

The fractions ⅔ and ⅝ must be reduced to a common denominator, before the one can be subtracted from the other.

⅔ = $\frac{14}{21}$ and ⅝ = $\frac{15}{21}$. Therefore

$$23\tfrac{2}{3} = 23\tfrac{14}{21}$$
$$8\tfrac{5}{7} = 8\tfrac{15}{21}$$

But $\frac{15}{21}$ is larger than $\frac{14}{21}$ and cannot be subtracted from it. To avoid this difficulty, 1 must be taken from 23 and reduced to 21ths, thus,

$$23\tfrac{14}{21} = 22 + 1\tfrac{14}{21} = 22\tfrac{35}{21}$$
$$8\tfrac{15}{21}$$
_____

Ans. 14$\frac{20}{21}$ yards.

$\frac{15}{21}$ taken from $\frac{35}{21}$ leaves $\frac{20}{21}$. Then 8 from 22 leaves 14. Ans. 14$\frac{20}{21}$ yards.

From the above examples it appears that in order *to add or subtract fractions, when they have a common denominator, we must add or subtract their numerators; and if they have not a common denominator, they must first be reduced to a common denominator.*

We find also the following rule to reduce them to a common denominator : *multiply all the denominators together, for a common denominator, and then multiply each numerator by all the denominators except its own.*

XX. This seems a proper place to introduce some contractions in division.

If 24 *barrels of flour cost* 192 *dollars, what is that a barrel ?*

This example may be performed by short division. First find the price of 6 barrels, and then of 1 barrel ; 6 barrels will cost ¼ of the price of 24 barrels.

192 (4
_____

Price of 6 bar.   48 (6
_____

· Price of 1 bar.   8 dolls. Ans.

If 56 *pieces of cloth cost* $7580.72, *what is it a piece ?*

First find the price of 7, or of 8 pieces, and then of 1 piece. 7 pieces will cost ⅛ of the price of 56 pieces.

$$7580.72 \ (8$$

Price of 7 pieces   947.59 (7

Price of 1 piece   **$135.37** Ans.

Divide $24674 equally among 63 men. How much will each have?

First find the share of 7 or 9 men, and then of 1 man. The share of 7 men will be ⅑ of the whole. The share of 9 men will be ⅐ of the whole.

$$24674 \ (9$$

Share of 7 men   2741⅔ (7

Share of 1 man   **391\frac{44}{63}$** Ans.

$$24674 \ (7$$

Share of 9 men   3524$\frac{6}{7}$ (9

Share of 1 man   **391\frac{44}{63}$** Ans.

In the first case I divide by 9, and then by 7. In dividing by 7 there is a remainder of 4⅔, which is $\frac{41}{9}$; this divided by 7 gives $\frac{44}{63}$. In the second case, I divide by 7 and then by 9. In dividing by 9 there is a remainder of 5$\frac{6}{7}$, which is $\frac{41}{7}$; this divided by 9 gives $\frac{44}{63}$ as before.

Divide 75345 dollars equally among 1800 men, how much will each have?

First find the share of 18 men, which will be $\frac{1}{100}$ part of the whole. $\frac{1}{100}$ part is found by cutting off the two right hand figures and making them the numerator of a fraction, thus, $753\frac{45}{100}$.

Share of 18 men $753\frac{45}{100}$ (18

    72

    ——   **41\frac{1}{2}\frac{45}{100}$** Ans. share of 1 man.

    33
    18

    ——

    15$\frac{45}{100}$ = $\frac{1545}{100}$; this divided by 18 is $\frac{1545}{1800}$.

It may be done as follows :

Share of 18 men $753\frac{45}{100}$ (6

Share of 3 men $125\frac{3\cdot45}{6\cdot100}$ (3

Share of 1 man $41\frac{1\cdot3\cdot45}{1\cdot6\cdot100}$ Ans.

In the last case I find the share of 3 men, and then of 1 man. In dividing by 6 there is a remainder $3\frac{45}{100}$, which is $\frac{3\cdot45}{100}$, this divided by 6 gives a fraction $\frac{3\cdot45}{6\cdot100}$. In dividing by 3 there is a remainder $2\frac{3\cdot45}{6\cdot100}$, which is equal to $\frac{1\cdot3\cdot45}{6\cdot100}$, this divided by 3 gives a fraction $\frac{1\cdot3\cdot45}{1\cdot6\cdot100}$, and the answer is $41\frac{1\cdot3\cdot45}{1\cdot6\cdot100}$ each.

From these examples we derive the following rule : *When the divisor is a compound number, separate the divisor into two or more factors, and divide the dividend by one factor of the divisor, and that quotient by another, and so on, until you have divided by the whole, and the last quotient will be the quotient required.*

When there are zeros at the right of the divisor, you may cut them off, and as many figures from the right of the dividend, making the figures so cut off the numerator of a fraction, and 1 with the zeros cut off, will be the denominator ; then divide by the remaining figures of the divisor.

XXI. In Art. XIX, it was observed, that if both the numerator and denominator of a fraction can be divided by the same number, without a remainder, it may be done, and the value of the fraction will remain the same. This gives rise to a question, how to find the divisors of numbers.

It is evident that if one number contain another a certain number of times, twice that number will contain the other twice as many times ; three times that number will contain the other thrice as many times, &c. that if one number is divisible by another, that number taken any number of times will be divisible by it also.

10 (and consequently any number of tens) is divisible by 2, 5, and 10 ; therefore if the right hand figure of any number is zero, the number may be divided by either 2, 5, or 10. If the right hand figure is divisible by 2, the number may be divided by 2. If the right hand figure is 5, the number may *be divided by '5·*

100 (and consequently any number of hundreds) is divisible by 4; therefore if the two right hand figures taken together are divisible by 4, the number may be divided by 4.

200 is divisible by 8; therefore if the hundreds are even, and the two right hand figures are divisible by 8, the number may be divided by 8. But if the hundreds are odd, it will be necessary to try the three right hand figures. 1000, being even hundreds, is divisible by 8.

To find if a number is divisible by 3 or 9, add together all the figures of the number, as if they were units, and if the sum is divisible by 3 or 9, the number may be divided by 3 or 9.

The number 387 is divisible by 3 or 9, because $3 + 8 + 7 = 18$, which is divisible by both 3 or 9.

The proof of the above rule is as follows: $10 = 9 + 1$; $20 = 2 \times 9 + 2$; $30 = 3 \times 9 + 3$; $52 = 5 \times 9 + 5 + 2$; $100 = 99 + 1$; $200 = 2 \times 99 + 2$; $387 = 3 \times 99 + 3 + 8 \times 9 + 8 + 7 = 3 \times 99 + 8 \times 9 + 3 + 8 + 7$. That is, in all cases, if a number of tens be divided by 9, the remainder will be equal to the number of tens; and if a number of hundreds be divided by 9, the remainder will always be equal to the number of hundreds. The same is true of thousands and higher numbers. Therefore, if the tens, hundreds, thousands, &c. of any number be divided separately by 9, the remainders will be the figures of that number, as in the above example 387. Now if the sum of these remainders be divisible by 9, the whole number must be so. But as far as the number may be divided by 9, it may be divided by 3; therefore, if the sum of the remainders, after dividing by 9, that is, the sum of the figures are divisible by 3, the whole number will be divisible by 3.

The numbers 615, 156, 3846, 28572 are divisible by 3, because the sum of the figures in the first is 12, in the second 12, in the third 21, and in the fourth 24.

The numbers 216, 378, 6453, and 804672 are divisible by 9, because the sum of the figures in the first is 9, in the second 18, in the third 18, and in the fourth 27.

When a number is divisible by both 2 and 3, it is divisible by their product 6. If it is divisible by 4 and 3 or 5 and 3, it is divisible by their products 12 and 15. In fine, when a number is divisible by any two or more numbers, it is divisible by their product.

N. B. To know if a number is divisible by 7, 11, 23, &c. it *must be found by trial.*

When two or more numbers can be divided by the same number without a remainder, that number is called their *common divisor*, and the greatest number which will divide them so, is called their *greatest common divisor.* When two or more numbers have several common divisors, it is evident that the greatest common divisor will be the product of them all.

In order to reduce a fraction to the lowest terms possible, it is necessary to divide the numerator and denominator by all their common divisors, or by their greatest common divisor at first.

Reduce $\frac{126}{342}$ to its lowest terms.

I observe in the first place that both numerator and denominator are divisible by 9, because the sum of the figures in each is 9. I observe also, that both are divisible by 2, because the right hand figure of each is so; therefore they are both divisible by 18. But it is most convenient to divide by them separately.

$$\frac{126}{342} \; (9 = \frac{14}{38} \; (2 = \frac{7}{19}.$$

7 and 19 have no common divisor, therefore $\frac{7}{19}$ cannot be reduced to lower terms.

The greatest common divisor cannot always be found by the above method. It will therefore be useful to find a rule by which it may always be discovered.

Let us take the same numbers 126 and 342.

126 is a number of even 18s, and 342 is a number of even 18s; therefore if 126 be subtracted from 342, the remainder 216 must be a number of even 18s. And if 126 be subtracted from 216, the remainder 90 must be a number of even 18s. Now I cannot subtract 126 from 90, but since 90 is a number of even 18s, if I subtract it from 126, the remainder 36 must be a number of even 18s. Now if 36 be subtracted from 90, the remainder 54 must be a number of even 18s. Subtracting 36 from 54, the remainder is 18. Thus by subtracting one number from the other, a smaller number was obtained every time, which was always a number of even 18s, until at last I came to 18 itself. If 18 be subtracted twice from 36 there will be no remainder. It is easy to see, that whatever be the common divisor, since each number is a certain number of times the common divisor, if one be subtracted from the other, the remainder will be a *certain number* of times the common divisor, that is, it will *have the same* divisor as the numbers themselves. And every

time the subtraction is made, a new number, smaller than the last, is obtained, which has the same divisor ; and at length the remainder must be the common divisor itself; and if this be subtracted from the last smaller number as many times as it can be, there will be no remainder. By this it may be known when the common divisor is found. It is the number which being subtracted leaves no remainder.

When one number is considerably larger than the other, division may be substituted for subtraction. The remainders only are to be noticed, no regard is to be paid to the quotient.

Reduce the fraction $\frac{330}{462}$ to its lowest terms.

Subtracting 330 from 462, there remains 132. 132 may be subtracted twice, or which is the same thing, is contained twice in 330, and there is 66 remainder. 66 may be subtracted twice from 132, or it is contained twice in 132, and leaves no remainder ; 66 therefore is the greatest common divisor. . Dividing both numerator and denominator by 66, the fraction is reduced to $\frac{5}{7}$.

Operation.

```
        462 (330        330 (66 = 5/7
        330 —           —
        — 1             462
330 (132
264 —
  — 2
132 (66
132 —
 — 2
...
```

From the above examples is derived the following general rule, to find the greatest common divisor of two numbers : *Divide the greater by the less, and if there is no remainder, that number is itself the divisor required; but if there is a remainder, divide the divisor by the remainder, and then divide the last divisor by that remainder, and so on, until there is no remainder, and the last divisor is the divisor required.*

If there be more than two numbers of which the greatest common divisor is to be found; find the greatest common divisor of two of them, and then take that common divisor and one of the other numbers, and find their greatest common divisor, and so on.

Reduce the fraction $\frac{9}{17}$ to its lowest terms.

```
      17 (9
       9 -
       — 1
```

```
       9 (8        1 is the greatest common divisor in
       8 —         this example.  Therefore the fraction
       — 1.        cannot be reduced.
      1 (1
      1 —
      – 1
      0
```

XXII. The method for finding the common denominator, given in Art. XIX. though always certain, is not always
the best; for it frequently happens that they may be reduced
to a common denominator, much smaller than the one obtained by that rule.

Reduce $\frac{5}{6}$ and $\frac{2}{9}$ to a common denominator.

According to the rule in Art. XIX., the common denomi-
nator will be 54, and $\frac{5}{6} = \frac{45}{54}$ and $\frac{2}{9} = \frac{12}{54}$.

It was observed Art. XIX., that the common denominator
may be any number, of which all the denominators are factors.   6 and 9 are both factors of 18, therefore they may be
both reduced to 18ths. $\frac{5}{6} = \frac{15}{18}$, and $\frac{2}{9} = \frac{4}{18}$.

When the fractions consist of small numbers, the least
denominator to which the fractions can be reduced, may be
easily discovered by trial ; but when they are large it is more
difficult.   It will, therefore, be useful to find a rule for it.

Any number, which is composed of two or more factors,
is called a *multiple* of any one of those factors.   Thus 18 is
a multiple of 2, or of 3, or of 6, or of 9.   It is also a *common* multiple of these numbers, that is, it may be produced
by multiplying either of them by some number.   The *least
common multiple* of two or more numbers, is the least number of which they are all factors.   54 is a common multiple
of 6 and 9, but their least common multiple is 18.

The least common denominator of two or more fractions
will be the least common multiple of all the denominators;
the fractions being previously reduced to their lowest terms.

One number will always be a multiple of another, when
the former contains all the factors of the latter. $6 = 2 \times 3$,
and $9 = 3 \times 3$, and $18 = 2 \times 3 \times 3$.   18 contains the factors 2 and 3 of 6, and 3 and 3 of 9. $54 = 2 \times 3 \times 3 \times 3.$

54, which is produced by multiplying 6 and 9, contains all these factors, and one of them, viz. 3, repeated. The reason why 3 is repeated is because it is a factor of both 6 and 9. By reason of this repetition, a number is produced 3 times as large as is necessary for the common multiple.

When the least common multiple of two or more numbers is to be found, if two or more of them have a common factor, it may be left out of all but one, because it will be sufficient that it enters once into the product.

These factors will enter once into the product, and only once, *if all the numbers which have common factors be divided by those factors; and then the undivided numbers, and these quotients be multiplied together, and the product multiplied by the common factors.*

If any of the quotients be found to have a common factor with either of the numbers, or with each other, they may be divided by that also.

Reduce $\frac{3}{4}$, $\frac{2}{9}$, $\frac{5}{6}$, and $\frac{4}{5}$, to the least common denominator.

The least common denominator will be the least common multiple of 4, 9, 6, and 5.

Divide 4 and 6 by 2, the quotients are 2 and 3. Then divide 3 and 9 by 3, the quotients are 1 and 3. Then multiplying these quotients, and the undivided number 5, we have $2 \times 1 \times 3 \times 5 = 30$. Then multiplying 30 by the two common factors 2 and 3, we have $30 \times 2 \times 3 = 180$, which is to be the common denominator.

Now to find how many 180ths each fraction is, take $\frac{3}{4}$, $\frac{2}{9}$, $\frac{5}{6}$, and $\frac{4}{5}$ of 180. Or observe the factors of which 180 was made up in the multiplication above. Thus $2 \times 1 \times 3 \times 5 \times 2 \times 3 = 180$. Then multiply the numerator of each fraction by the numbers by which the factors of its denominator were multiplied.

The factors 2 and 2 of the denominator of the first fraction, were multiplied by 1, 3, 3, and 5. The factors 3 and 3, of the second, were multiplied by 2, 1, 5, and 2. The factors 2 and 3, of the third, were multiplied by 2, 1, 3, 5; and 5, the denominator of the fourth, was multiplied by 2, 2, 1, 3, and 3.

$$\frac{3}{4} = \frac{135}{180}; \quad \frac{2}{9} = \frac{40}{180}; \quad \frac{5}{6} = \frac{148}{180}; \quad \frac{4}{5} = \frac{144}{180}.$$

XXIII. *If a horse will eat $\frac{1}{3}$ of a bushel of oats in a day, how long will 12 bushels last him?*

In this question it is required to find how many times $\frac{1}{3}$

16

of a bushel is contained in 12 bushels. In 12 there are $\frac{36}{3}$, therefore 12 bushels will last 36 days.

At ¼ of a dollar a bushel, how many bushels of corn may be bought for 15 dollars?

First find how many bushels might be bought at $\frac{1}{5}$ of a dollar a bushel. It is evident, that each dollar would buy 5 bushels; therefore 15 dollars would buy 15 times 5, that is, 75 bushels. But since it is $\frac{3}{5}$ instead of $\frac{1}{5}$ of a dollar a bushel, it will buy only $\frac{1}{3}$ as much, that is, 25 bushels.

This question is to find how many times $\frac{3}{5}$ of a dollar, are contained in 15 dollars. It is evident, that 15 must be reduced to 5ths, and then divided by 3.

$$
\begin{array}{r}
15 \\
5 \\
\hline
75 \ (3 \\
\hline
25 \text{ bushels.}
\end{array}
$$

The above question is on the same principle as the following.

How much corn, at 5 shillings a bushel, may be bought for 23 dollars?

The dollars in this example must be reduced to shillings, before we can find how many times 5 shillings are contained in them; that is, they must be reduced to 6ths, before we can find how many times $\frac{5}{6}$ are contained in them.

$$
\begin{array}{r}
23 \\
6 \\
\hline
138 \ (5 \\
\hline
\end{array}
$$

Ans. $27\frac{3}{5}$ bushels.

$23 = \frac{138}{6}$ and $\frac{5}{6}$ are contained $27\frac{3}{5}$ times in $\frac{138}{6}$.

If $7\frac{3}{8}$ yds. of cloth will make 1 suit of clothes, how many suits will 48 yards make?

If the question was given in yards and quarters, it is evident both numbers must be reduced to quarters. In this instance then, they must be reduced to 8ths.

$$
7\frac{3}{8} = \frac{59}{8} \text{ and } 48 = \frac{384}{8}.
$$
$$
\begin{array}{r}
384 \ (59 \\
354 \ \overline{} \\
\hline
30
\end{array}
\quad 6\frac{30}{11} \text{ suits. Ans.}
$$

In the three last examples, the purpose is to find how many times a fraction is contained in a whole number. This is dividing a whole number by a fraction, for which we find the following rule : *Reduce the dividend to the same denomination as the divisor, and then divide by the numerator of the fraction.*

Note. If the divisor is a mixed number, it must be reduced to an improper fraction.

N. B. The above rule amounts to this ; *multiply the dividend by the denominator of the divisor, and then divide it by the numerator.*

At $\frac{1}{4}$ of a dollar a bushel, how many bushels of potatoes may be bought for $\frac{3}{4}$ of a dollar.

$\frac{1}{4}$ is contained in $\frac{3}{4}$ as many times as 1 is contained in 3. Ans. 3 bushels.

If $\frac{3}{10}$ of a ton of hay will keep a horse 1 month, how many horses will $\frac{9}{10}$ of a ton keep the same time?

$\frac{3}{10}$ are contained in $\frac{9}{10}$ as many times as 3 are contained in 9. Ans. 3 horses.

At $\frac{1}{5}$ of a dollar a pound, how many pounds of figs may be bought for $\frac{3}{4}$ of a dollar?

5ths and 4ths are different denominations ; before one can be divided by the other, they must be reduced to the same denomination ; that is, reduced to a common denominator.

$\frac{1}{5} = \frac{4}{20}$ and $\frac{3}{4} = \frac{15}{20}$. $\frac{4}{20}$ are contained in $\frac{15}{20}$ as many times as 4 are contained in 15. Ans. $3\frac{3}{4}$ lb.

At $7\frac{3}{5}$ dolls. a yard, how many yards of cloth may be bought for $57\frac{5}{8}$ dollars?

$7\frac{3}{5} = \frac{38}{5}$ and $57\frac{5}{8} = \frac{461}{8}$. 5ths and 8ths are different denominations ; they must, therefore, be reduced to a common denominator.

$\frac{38}{5} = \frac{304}{40}$ and $\frac{461}{8} = \frac{2305}{40}$.

$$\begin{array}{r} 2305\ (304 \\ 2128 \quad\underline{} \\ \underline{}\quad 7\frac{177}{304}\ \text{yards.} \\ 177 \end{array}$$

From the above examples we deduce the following rule, for dividing one fraction by another :

If the fractions are of the same denomination, divide the numerator of the dividend by the numerator of the divisor.

If the fractions are of different denominations, they must first be reduced to a common denominator.

If either or both of the numbers are mixed numbers, they must first be reduced to improper fractions.

Note. As the common denominator itself is not used in the operation, it is not necessary actually to find it, but only to multiply the numerators by the proper numbers to reduce them. By examining the above examples, it will be found that this purpose is effected, *by multiplying the numerator of the dividend by the denominator of the divisor, and the denominator of the dividend by the numerator of the divisor.* Thus in the third example; multiplying the numerator of $\frac{3}{4}$ by 5 and the denominator by 1, it becomes $\frac{15}{4}$, which reduced is $3\frac{3}{4}$ pounds as before.

XXIV. *A owned $\frac{1}{5}$ of a ticket, which drew a prize. A's share of the money was 567 dollars. What was the whole prize?*

$\frac{1}{5}$ of a number make the whole number. Therefore the whole prize was 5 times A's share.

$$567$$
$$5$$
$$\overline{}$$

Ans. 2835 dollars.

A man bought $\frac{1}{7}$ of a ton of iron for $13\frac{5}{8}$ dollars; what was it a ton?

$\frac{7}{7}$ make the whole, therefore the whole ton cost 7 times $13\frac{5}{8}$.

$$13\frac{5}{8}$$
$$7$$
$$\overline{}$$

Ans. $95\frac{3}{8}$ dolls.

A man bought $\frac{5}{13}$ of a ton of iron for 40 dollars; what was it a ton?

$\frac{5}{13}$ are 5 times as much as $\frac{1}{13}$. If $\frac{5}{13}$ cost 40 dollars, $\frac{1}{13}$ must cost $\frac{1}{5}$ of 40. $\frac{1}{5}$ of 40 is 8, and 8 is $\frac{1}{12}$ of 96. Ans. 96 dollars.

A man bought $\frac{2}{3}$ of a ton of hay for 17 dollars; what was it a ton?

$\frac{2}{3}$ are 3 times as much as $\frac{1}{3}$. Since $\frac{2}{3}$ cost 17 dollars, $\frac{1}{3}$ must cost $\frac{1}{2}$ of 17, and $\frac{3}{3}$ must cost $\frac{3}{2}$ of 17.

17 (3	or multiplying first by	17
———	the denominator	5
$5\frac{2}{3}$		——
5		85 (3
——		——

Ans. $28\frac{1}{3}$ dolls. $28\frac{1}{3}$ dolls.

If $4\frac{3}{5}$ *firkins of butter cost* 33 *dollars, what is that a fir kin ?*

$4\frac{3}{5} = \frac{23}{5}$. First we must find what $\frac{1}{5}$ costs. $\frac{1}{5}$ is $\frac{1}{23}$ part of $\frac{23}{5}$, therefore $\frac{1}{5}$ will cost $\frac{1}{23}$ of 33 dollars, and $\frac{5}{5}$ will cost $\frac{5}{23}$ of 33 dollars.

$$\begin{array}{r} 33 \\ 5 \\ \hline 165 \ (22 \\ 154 \ \underline{\quad} \\ \hline 11 \end{array} \quad 7\frac{11}{22} = 7\frac{1}{2} \text{ dollars.}$$

The six last examples are evidently of the same kind. In all of them a part or several parts of a number were given to find the whole number. They are exactly the reverse of the examples in Art. XVI. If we examine them still farther, we shall find them to be division. In the last example, if 4 firkins instead of $4\frac{3}{5}$ had been given, it would evidently be division ; as it is, the principle is the same. It is therefore dividing a whole number by a fraction ; the general rule is, *to find the value of one part, and then of the whole.* *To find the value of one part, divide the dividend by the nume-rator of the divisor ; and then to find the whole number, multiply the part by the denominator.*

Or, according to the two last examples, *multiply the divi-dend by the denominator of the divisor, and divide by the numerator.*

N. B. This last rule is the same as that in Art. XXIII. This also shows this operation to be division.

Note. If the divisor is a mixed number, reduce it to an improper fraction.

If $\frac{1}{5}$ *of a yard of cloth cost* $\frac{3}{7}$ *of a dollar, what will a yard cost ?*

It is evident that the whole yard will cost 5 times $\frac{3}{7}$, which is $\frac{15}{7} = 2\frac{1}{7}$ dollars.

If $\frac{3}{7}$ *of a yard of cloth cost* $\frac{5}{8}$ *of a dollar, what is that a yard ?*

If $\frac{3}{7}$ cost $\frac{5}{8}$, $\frac{1}{7}$ must cost $\frac{1}{3}$ of $\frac{5}{8}$; $\frac{1}{3}$ of $\frac{5}{8}$ is $\frac{5}{24}$; $\frac{5}{24}$ being $\frac{1}{7}$, 7 times $\frac{5}{24}$ or $\frac{35}{24} = 1\frac{11}{24}$ dollars must be the price of a yard.

If 3⅛ barrels of flour cost 23¾ dollars, what is that a barrel?

3⅛ = ²⁹⁄₈ and 23¾ = ¹⁶³⁄₇. If ²⁹⁄₈ of a barrel cost ¹⁶³⁄₇ of a dollar, ⅛ of a barrel will cost ¹⁄₂₉ of ¹⁶³⁄₇. ¹⁄₂₉ of ¹⁶³⁄₇ is ¹⁶³⁄₂₀₃. ¹⁶³⁄₂₀₃ being ⅛ of the price of 1 barrel, 8 times ¹⁶³⁄₂₀₃ will be the price of a barrel. 8 times ¹⁶³⁄₂₀₃ = ¹³⁰⁴⁄₂₀₃ = 6⁸⁶⁄₂₀₃ dollars. Ans. 6⁸⁶⁄₂₀₃ dollars per barrel.

The three last examples are of the same kind as those which precede them ; the only difference is, that in these, the part which is given, or the dividend, is a fraction or mixed number.

In this case the dividend, if a mixed number, must be reduced to an improper fraction ; then in order to divide the dividend by the numerator of the divisor, it will generally be necessary to multiply the denominator of the dividend by the numerator of the divisor.

From this article and the preceding, we derive the following general rule, to divide by a fraction, whether the dividend be a whole number or not : *Multiply the dividend by the denominator of the divisor, and divide the product by the numerator. If the divisor is a mixed number, it must be changed to an improper fraction.*

———◆———

DECIMAL FRACTIONS.

XXV. We have seen that the nine digits may be made to express different values, by putting them in different places, and that any number, however large, may be expressed by them. We shall now see how they may be made to express numbers less than unity, (that is, fractions,) in the same manner as they do those larger than unity.

Suppose the unit to be divided into ten equal parts. These are called tenths, and ten of them make 1, in the same manner as ten units make 1 ten, and as ten tens make 1 hundred, &c. In the common way, 3 tenths is written ³⁄₁₀, and 47 and 3 tenths is written 47³⁄₁₀. Now if we assign a place for tenths, as we do for units, tens, &c. it is evident that they may be written without the denominator, and they *will be always* understood as tenths. It is agreed to write *tenths* at the right hand of the units, separated from them a point (.). Hitherto we have been accustomed to consider the right hand figure as expressing units ; we still consider units as the starting point, and must therefore make a mark, in order to show which we intend for units. Thus $47\frac{3}{10}$. 47 signifies 4 tens and 7 units ; then if we wish to write $\frac{3}{10}$, we make a point at the right of 7, and then write 3, thus, 47.3. This is read forty-seven and three tenths.

Again, suppose each tenth to be divided into ten equal parts : the whole unit will then be divided into one hundred equal parts. But they were made by dividing tenths into ten equal parts, therefore ten hundredths will make one tenth. Hundredths then may with propriety be written at the right of tenths, but there is no need of a mark to distinguish these, for the place of units being the starting point, when that is known, all the others may be easily known.

$7\frac{4}{100}$ is written 7.04. 83.57 is read 83 and $\frac{5}{10}$ and $\frac{7}{100}$, or since $\frac{5}{10} = \frac{50}{100}$ we may read it $83\frac{57}{100}$, which is a shorter expression.

Again, suppose each hundredth to be divided into ten equal parts ; these will be thousandths. And since ten of the thousandths make one hundredth, these may with propriety occupy the place at the right of the hundredths, or the third place from the units.

It is easy to see that this division may be carried as far as we please. The figures in each place at the right, signifying parts 1 tenth part as large as those in the one at the left of it.

Beginning at the place of units and proceeding towards the left, the value of the places increases in a tenfold proportion, and towards the right it diminishes in a tenfold proportion.

Fractions of this kind may be written in this manner, when there are no whole numbers to be written with them. $\frac{4}{10}$ for example may be written 0.4, or simply .4. $\frac{3}{100}$ may be written 0.03 or .03. $\frac{87}{100}$ may be written .87. The point always shows where the decimals begin. Since the value of a figure depends entirely upon the place in which it is written, great care must be taken to put every one in its proper place.

Fractions written in this way are called *decimal* fractions, from the Latin word *decem*, which signifies ten, because they increase and diminish in a tenfold proportion.

It is *important* to remark that $\frac{3}{10} = \frac{30}{100} = \frac{300}{1000} = \frac{3000}{10000}$.

&c. and that $\frac{5}{100} = \frac{50}{1000} = \frac{500}{10000}$, &c. and $\frac{7}{1000} = \frac{70}{10000}$, consequently $\frac{3}{10} + \frac{5}{100} + \frac{7}{1000} + \frac{2}{10000} = \frac{3572}{10000} = 0.3572$.
Any other numbers may be expressed in the same manner.
From this it appears that any decimal may be reduced to a lower denomination, simply by annexing zeros. Also any number of decimal figures may be read together as whole numbers, giving the name of the lowest denomination to the whole.

Thus 0.38752 is actually $\frac{3}{10} + \frac{8}{100} + \frac{7}{1000} + \frac{5}{10000} + \frac{2}{100000}$, but it may all be read together $\frac{38752}{100000}$, thirty-eight thousand, seven hundred and fifty-two hundred-thousandths. Any whole number may be reduced to tenths, hundredths, &c. by annexing zeros. 27 is 270 tenths, 2700 hundredths, &c. consequently 27.35 may be read two thousand, seven hundred and thirty-five hundredths, $\frac{2735}{100}$. In like manner any whole number and decimal may be read together, giving it the name of the lowest denomination. It is evident that a zero at the right of decimals does not alter the value, but a zero at the left diminishes the value tenfold.

It is evident that any decimal may be changed to a common fraction, by writing the denominator, which is always understood, under the fraction. Thus .75 may be written $\frac{75}{100}$, then reducing it to its lowest terms it becomes $\frac{3}{4}$. The denominator will always be 1, with as many zeros as there are decimal places, that is, one zero for tenths, two for hundredths &c.

The following table exhibits the places with their names, as far as ten-millionths, together with some examples.

| | | | | | | | | | | Thousands. | Hundreds. | Tens. | Units. | Tenths. | Hundredths. | Thousandths. | Ten-thousandths. | Hundred-thousandths. | Mills. | Ten-millionths. |

6 and 7 tenths $6\frac{7}{10}$ . . . 6 .7 . . . . . .

44 and 3 hundredths $44\frac{3}{100}$ . . 4 4 .0 3 . . . . .

50 and 64 hundredths $50\frac{64}{100}$ . . 5 0 .6 4 . . . . . .

243 and 87 thousandths $243\frac{87}{1000}$. 2 4 3 .0 8 7 . . . . .

9247 and 204 thousandths

$9247\frac{204}{1000}$ 9 2 4 7 .2 0 4 . . . .

42 and 7 ten-thousandths

$42\frac{7}{10000}$ . . 4 2 .0 0 0 7 . . .

3 and 904 ten-thousandths

$3\frac{904}{10000}$ . . . 3 .0 9 0 4 . . .

9 tenths $\frac{9}{10}$ . . . .9 . . . . . .

3 thousandths $\frac{3}{1000}$ . . . .0 0 3 . . . .

29 hundredths $\frac{29}{100}$ . . . .2 9 . . . . .

8 hundred-thousandths $\frac{8}{100000}$ . . . .0 0 0 0 8 . .

67 millionths $\frac{67}{1000000}$ . . . .0 0 0 0 6 7 .

3064 ten-millionths $\frac{3064}{10000000}$ 0 0 0 3 0 6 4

In Federal money the parts of a dollar are adapted to the decimal division ot the unit. The dollar being the unit. dimes are tenths, cents are hundredths, and mills are thousandths.

For example, 25 dollars, 8 dimes, 3 cents, 7 mills, are written $25.837, that is, $25\frac{837}{1000}$ dollars.

XXVI. *A man purchased a cord of wood for 7 dollars, 3 dimes, 7 cents, 5 mills, that is, $7.375 ; a gallon of molasses for $0.43 ; 1 lb. of coffee for $0.27 ; a firkin of butter for $8 ; a gallon of brandy for $0.875 ; and 4 eggs for $0.03. How much did they all come to ?*

It is easy to see that dollars must be added to dollars,

dimes to dimes, cents to cents, and mills to mills. They may be written down thus:

$7.375
0.430
0.270
8.000
0.875
0.030

Ans. $16.980

A man bought $3\frac{3}{10}$ *barrels of flour at one time,* $8\frac{63}{100}$ *barrels at another,* $\frac{873}{1000}$ *barrel at a third, and* $15\frac{784}{1000}$ *at a fourth. How many barrels did he buy in the whole?*

These may be written without the denominators, as follows: 3.3 barrels, 8.63 barrels, .873 barrel, 15.784 barrels. It is evident that units must be added to units, tenths to tenths, &c. For this it may be convenient to write them down so that units may stand under units, tenths under tenths, &c. as follows:

3.3
8.63
 .873
15.784

Ans. 28.587 barrels. That is, $28\frac{587}{1000}$ barrels.

I say 3 (thousandths) and 4 (thousandths) are 7 (thousandths,) which I write in the thousandths' place. Then 3 (hundredths) and 7 (hundredths) are 10 (hundredths) and 8 (hundredths) are 18 (hundredths,) that is, 1 tenth and 8 hundredths. I reserve the 1 tenth and write the 8 hundredths in the hundredths' place. Then 1 tenth (which was reserved) and 3 tenths are 4 tenths, and 6 are 10, and 8 are 18, and 7 are 25 (tenths,) which are 2 whole ones and 5 tenths. I reserve the 2 and write the 5 tenths in the tenths' place. Then 2 (which were reserved) and 3 are 5, and 8 are 13, and 5 are 18, which is 1 ten and 8. I write the 8 and carry the 1 ten to the 1 ten, which makes 2 tens. The answer is 28.587 barrels.

It appears that *addition of decimals is performed in precisely the same manner as addition of whole numbers. Care must be taken to add units to units, tenths to tenths, &c. To prevent mistakes it will generally be most convenient to write them, so that units may stand under units, tenths under tenths, &c.*

It is plain that the operations on decimal fractions are as easy as those on whole numbers, but fractions of this kind do not often occur. We shall now see that common fractions may be changed to decimals.

A merchant bought 6 pieces of cloth ; the first containing 14½ yards, the second 37⅗, the third 4¼, the fourth 17¾, the fifth 19⅜, and the sixth 42¹³⁄₂₀. How many yards in the whole ?

$$14\tfrac{1}{2}$$
$$37\tfrac{3}{5}$$
$$4\tfrac{1}{4}$$
$$17\tfrac{3}{4}$$
$$19\tfrac{3}{8}$$
$$42\tfrac{13}{20}$$

To add these fractions together in the common way, they must be reduced to a common denominator. But instead of reducing them to a common denominator in the usual way, we may reduce them to decimals, which is in fact reducing them to a common denominator; but the denominator is of a peculiar kind.

$\frac{1}{2} = \frac{5}{10}, \frac{3}{5} = \frac{6}{10}.$ $\frac{1}{4}$ cannot be changed to tenths, but it may be changed to hundredths. $\frac{1}{4} = \frac{25}{100}, \frac{3}{4} = \frac{75}{100}.$ $\frac{3}{8}$ cannot be changed to hundredths, but it may be changed to thousandths. $\frac{3}{8} = \frac{375}{1000}.$ $\frac{13}{20}$ may be reduced to hundredths. $\frac{1}{20} = \frac{5}{100}$, and $\frac{13}{20} = \frac{65}{100}.$

Writing the fractions now without their denominators in the form of decimals, they become

$$14.5$$
$$37.6$$
$$4.25$$
$$17.75$$
$$19.375$$
$$42.65$$

Ans. 136.125 yards or $136\frac{125}{1000} = 136\frac{1}{8}$ yards.

Common fractions cannot always be changed to decimals so easily as those in the above example, but since there will be frequent occasion to change them, it is necessary to find a principle, by which it may always be done.

A man divided 5 bushels of wheat equally among 8 persons ; how much did he give them apiece ?

He gave them $\frac{1}{8}$ of a bushel apiece, expressed in the form of common fractions; but it is proposed to express it in decimals.

I first suppose each bushel to be divided into 10 equal parts or tenths. The five bushels make $\frac{5 0}{1 0}$. I perceive that I cannot divide $\frac{5 0}{1 0}$ into exactly 8 parts, therefore I suppose each of these parts to be divided into 10 equal parts; these parts will be hundredths. $5 = \frac{5 0 0}{1 0 0}$. But 500 cannot be divided by 8 exactly, therefore I suppose these parts to be divided again into 10 parts each. These parts will be thousandths. $5 = \frac{5 0 0 0}{1 0 0 0}$. 5000 may be divided by 8 exactly, $\frac{1}{8}$ of $\frac{5 0 0 0}{1 0 0 0}$ is $\frac{6 2 5}{1 0 0 0}$, or .625. Ans. .625 of a bushel each.

Instead of trying until I find a number that may be exactly divided, I can perform the work as I make the trials. For instance, I say 5 bushels are equal to $\frac{5 0}{1 0}$ of a bushel. $\frac{1}{8}$ of $\frac{5 0}{1 0}$ is $\frac{6}{1 0}$, and there are $\frac{2}{1 0}$ left to be divided into 8 parts. I then suppose these 2 tenths to be divided into ten equal parts each. They will make 20 parts, and the parts are hundredths. $\frac{1}{8}$ of $\frac{2 0}{1 0 0}$ are $\frac{2}{1 0 0}$, and there are $\frac{4}{1 0 0}$ left to be divided into 8 parts. I suppose these 4 hundredths to be divided into 10 parts each. They will make 40 parts, and the parts will be thousandths. $\frac{1}{8}$ of $\frac{4 0}{1 0 0 0}$ is $\frac{5}{1 0 0 0}$. Bringing the parts $\frac{6}{1 0}$, $\frac{2}{1 0 0}$, and $\frac{5}{1 0 0 0}$ together, they make $\frac{6 2 5}{1 0 0 0}$ or .625 of a bushel each, as before.

The operation may be performed as follows:

```
      50 (8
      48 ———
      ——— .625
      20
      16
      ——
         40
         40
         ——
         ..
```

I write the 5 as a dividend and the 8 as a divisor. Then I multiply 5 by 10, (that is, I annex a zero) in order to reduce the 5 to tenths. Then $\frac{1}{8}$ of 50 is 6, which I write in the quotient and place a point before it, because it is tenths. There is 2 remainder. I multiply the 2 by 10, in order to *reduce it to* hundredths. $\frac{1}{8}$ of 20 is 2, and there is 4 remainder. I multiply the 4 by 10, in order to reduce it to thousandths. $\frac{1}{8}$ of 40 is 5. The answer is .625 bushels each, as before.

In Art. X. it was shown, that when there is a remainder after division, in order to complete the quotient, it must be written over the divisor, and annexed to the quotient. This fraction may be reduced to a decimal, by annexing zeros, and continuing the division.

Divide 57 barrels of flour equally among 16 men.

$$
\begin{array}{r}
57\ (16 \\
48 \quad \overline{} \\
\overline{}\ 3.5625\ \text{barrels each.} \\
90 \\
80 \\
\overline{} \\
100 \\
96 \\
\overline{} \\
40 \\
32 \\
\overline{} \\
80 \\
80 \\
\overline{} \\
\cdot\ \cdot
\end{array}
$$

In this example the answer, according to Art. X., is $3\frac{9}{16}$ bushels. But instead of expressing it so, I annex a zero to the remainder 9, which reduces it to tenths, then dividing, I obtain 5 tenths to put into the quotient, and I separate it from the 3 by a point. There is now a remainder 10, which I reduce to hundredths, by annexing a zero. And then I divide again, and so on, until there is no remainder.

The first remainder is 9, this is 9 bushels, which is yet to be divided among the 16 persons; when I annex a zero I reduce it to tenths. The second remainder 10 is so many tenths of a bushel, which is yet to be divided among the 16 persons. When I annex a zero to this I reduce it to hundredths. The next remainder is 4 hundredths, which is yet to be divided. By annexing a zero to this it is reduced to thousandths, and so on.

The division in this example stops at ten-thousandths; the reason is, because 10000 is exactly divisible by 16. If I take $\frac{9}{16}$ of $\frac{10000}{10000}$ I obtain $\frac{5625}{10000}$, or .5625, as above.

There are many common fractions which require so many

figures to express their value exactly in decimals, as to render them very inconvenient. There are many also, the value of which cannot be exactly expressed in decimals. In most calculations, however, it will be sufficient to use an approximate value. The degree of approximation necessary, must always be determined by the nature of the case. For example, in making out a single sum of money, it is considered sufficiently exact if it is right within something less than 1 cent, that is, within less than $\frac{1}{100}$ of a dollar. But if several sums are to be put together, or if a sum is to be multiplied, mills or thousandths of a dollar must be taken into the account, and sometimes tenths of mills or ten-thousandths. In general, in questions of business, three or four decimal places will be sufficiently exact. And even where very great exactness is required, it is not very often necessary to use more than six or seven decimal places.

A merchant bought 4 pieces of cloth; the first contained $28\frac{3}{5}$ *yards; the second* $34\frac{2}{7}$; *the third* $30\frac{1}{15}$; *and the fourth* $42\frac{7}{9}$ *yards. How many yards in the whole?*

In reducing these fractions to decimals, they will be sufficiently exact if we stop at hundredths, since $\frac{1}{100}$ of a yard is only about $\frac{1}{3}$ of an inch.

30 (5	200 (7	100 (15	700 (9
.6	.28 +	.07 —	.78 —

$\frac{3}{5}$ is exactly .6. If we were to continue the division of $\frac{2}{7}$, it would be .28571, &c.; in fact it would never terminate; but .28 is within about one $\frac{1}{2}$ of $\frac{1}{100}$ of a yard, therefore sufficiently exact. $\frac{1}{15}$ is not so much as $\frac{1}{10}$, therefore the first figure is in the hundredths' place. The true value is .0666, &c., but because $\frac{6}{1000}$ is more than $\frac{1}{2}$ of $\frac{1}{100}$, I call it .07 instead of .06. $\frac{7}{9}$ is equal to .7777, &c. This would never terminate. Its value is nearer .78 than .77, therefore I use .78.

When the decimal used is smaller than the true one, it is well to make the mark + after it, to show that something more should be added, as $\frac{2}{7} = .28 +$. When the fraction is too large the mark — should be made to show that *something should* be subtracted, as $\frac{1}{15} = .07$ —.

The numbers to be added will now stand thus:

$$28\tfrac{3}{5} = 28.60$$
$$34\tfrac{2}{7} = 34.28 +$$
$$30\tfrac{1}{15} = 30.07 -$$
$$42\tfrac{7}{9} = 42.78 -$$

Ans. 135.75 yards, or $135\tfrac{75}{100} = 135\tfrac{3}{4}$.

From the above observations we obtain the following general rule for changing a common fraction to a decimal: *Annex a zero to the numerator, and divide it by the denominator, and then if there be a remainder, annex another zero, and divide again, and so on, until there is no remainder, or until a fraction is obtained, which is sufficiently exact for the purpose required.*

Note. When one zero is annexed, the quotient will be tenths, when two zeros are annexed, the quotient will be hundredths, and so on. Therefore, if when one zero is annexed, the dividend is not so large as the divisor, a zero must be put in the quotient with a point before it, and in the same manner after two or more zeros are annexed, if it is not yet divisible, as many zeros must be placed in the quotient.

Two men talking of their ages, one said he was $37\tfrac{3847}{14783}$ years old, and the other said he was $64\tfrac{213}{250}$ years old. What was the difference of their ages?

If it is required to find an answer within 1 minute, it will be necessary to continue the decimals to seven places, for 1 minute is $\tfrac{1}{525600}$ of a year. If the answer is required only within hours, five places are sufficient; if only within days, four places are sufficient.

$$64\tfrac{213}{250} = 64.8520000$$
$$37\tfrac{3847}{14783} = 37.2602313 +$$

Ans. 27.5917687 years.

It is evident that units must be subtracted from units, tenths from tenths, &c. If the decimal places in the two numbers are not alike, they may be made alike by annexing zeros. *After the numbers are prepared, subtraction is performed precisely as in whole numbers.*

Multiplication of Decimals.

XXVII. *How many yards of cloth are there in seven pieces, each piece containing* 19⅞ *yards*?

$$19\tfrac{7}{8} = 19.875$$
$$7$$

Ans. 139.125 = $139\tfrac{125}{1000}$ = 139⅛ yards.

N. B. All the operations on decimals are performed in precisely the same manner as whole numbers. All the difficulty consists in finding where the separatrix, or decimal point, is to be placed. This is of the utmost importance, since if an error of a single place be made in this, their value is rendered ten times too large or ten times too small. The purpose of this article and the next is to show where the point must be placed in multiplying and dividing.

In the above example there are decimals in the multiplicand, but none in the multiplier. It is evident from what we have seen in adding and subtracting decimals, that in this case there must be as many decimal places in the product, as there are in the multiplicand. It may perhaps be more satisfactory if we analyze it.

7 times 5 thousandths are 35 thousandths, that is, 3 hundredths and 5 thousandths. Reserving the hundredths, I write the 5 thousandths. Then 7 times 7 hundredths are 49 hundredths, and 3 (which I reserved) are 52 hundredths, that is, 5 tenths and 2 hundredths. I write the two hundredths, reserving the 5 tenths. Then 7 times 8 tenths are 56 tenths, and 5 (which I reserved) are 61 tenths, that is, 6 whole ones and 1 tenth. I write the 1 tenth, reserving the 6 units. Then 7 times 9 are 63, and 6 are 69, &c. It is evident then, that there must be thousandths in the product; as there are in the multiplicand. The point must be made between the third and fourth figure from the right, as in the multiplicand, and the answer will stand thus, 139.125 yards.

Rule. When there are decimal figures in the multiplicand only, cut off as many places from the right of the product for decimals, as there are in the multiplicand.

If a ship is worth 24683 *dollars, what is a man's share worth, who owns* ⅜ *of her.*

⅜ = .375 = $\tfrac{375}{1000}$. The question then is, to find $\tfrac{375}{1000}$ of

24683 dollars. First find $\frac{1}{1000}$ of it, that is, divide it by 1000.- This is done by cutting off three places from the right (Art. XI.) thus 24.683, that is, $24\frac{683}{1000}$, because 683 is a remainder and must be written over the divisor. In fact it is evident that $\frac{1}{1000}$ of 24683 is $\frac{24683}{1000} = 24\frac{683}{1000}$. But since this fraction is thousandths, it may stand in the form of a decimal, thus 24.683.

It is a general rule then, *that when we divide by 10, 100, 1000, &c. which is done by cutting off figures from the right, the figures so cut off may stand as decimals, because they will always be tenths, hundredths, &c.*

$\frac{1}{1000}$ of 24683 then is 24.683 and $\frac{375}{1000}$ of it will be 375 times 24.683. Therefore 24.683 must be multiplied by 375.

24.683	24683
375	.375
123415	123415
172781	172781
74049	74049
$9256.125 Ans.	$ 9256.125

This result must have three decimal places, because tho multiplicand has three. The answer is 9256 dollars, 12 cents, and 5 mills. But the purpose was to multiply 24683 by .375, in which case the multiplier has three decimal places, and the multiplicand none. We pointed off as many places from the right of the multiplicand, as there were in the multiplier, and then used the multiplier as a whole number. This in fact makes the same number of decimal places in the product as there are in the multiplier.

We may arrive at this result by another mode of reasoning. Units multiplied by tenths will produce tenths; units multiplied by hundredths will produce hundredths; units multiplied by thousandths will produce thousandths, &c.

In the second operation of the above example, observe, that .375 is $\frac{3}{10}$, and $\frac{7}{100}$, and $\frac{5}{1000}$, then $\frac{1}{1000}$ of 3 is $\frac{3}{1000}$, and $\frac{5}{1000}$ of 3 is $\frac{15}{1000}$, which is $\frac{1}{100}$ and $\frac{5}{1000}$, set down the 5 thousandths in the place of thousandths, reserving the $\frac{1}{100}$. Then $\frac{1}{1000}$ of 80 is $\frac{80}{1000}$, or $\frac{8}{100}$, and 5 times $\frac{8}{100}$ is $\frac{40}{100}$, and $\frac{1}{100}$ (which was reserved) are $\frac{41}{100}$, equal to $\frac{4}{10}$ and $\frac{1}{100}$. Set down the $\frac{1}{100}$ in the hundredth's place, &c. This shows also, *that when there are no decimals in the multiplicand,*

17 *

there must be as many decimal places in the product as in the multiplier.

It was observed that when a whole number is to be multiplied by 10, 100, &c. it is done by annexing as many zeros to the right of the number as there are in the multiplier, and to divide by these numbers, it is done by cutting off as many places as there are zeros in the divisor. When a number containing decimals is to be multiplied or divided by 10, 100, &c. it is done by removing the decimal point as many places to the right for multiplication, and to the left for division, as there are zeros in the multiplier or divisor. If, for example, we wish to multiply 384.785 by 10, we remove the point one place to the right, thus, 3847.85, if by 100, we remove it two places, thus, 38478.5. If we wish to divide the same number by 10, we remove the point one place to the left, thus, 38.4785 ; if by 100, we remove it two places, thus, 3.84785. The reason is evident, for removing the point one place towards the right, units become tens, and the the tenths become units, and each figure in the number is increased tenfold, and when removed the other way each figure is diminished tenfold, &c.

How much cotton is there in $3\frac{7}{10}$ bales, each bale containing $4\frac{3}{4}$ cwt.

$$3\frac{7}{10} = 3.7 \; ; \; 4\frac{3}{4} = 4.75.$$

In this example there are decimals in both multiplicand and multiplier.

$$\begin{array}{r} 4.75 \\ 3.7 \\ \hline 3325 \\ 1425 \\ \hline \end{array}$$

Ans. 17.575 cwt.

3.7 is the same as $\frac{37}{10}$, we have to find $\frac{37}{10}$ of 4.75. Now $\frac{1}{10}$ of 4.75, we have just seen, must be .475, and $\frac{37}{10}$ is 37 times as much. We must therefore multiply .475 by 37, which gives 17.575 cwt.

We shall obtain the same result if we express the whole in the form of common fractions. $4.75 = 4\frac{75}{100} = \frac{475}{100}$, and $3.7 = \frac{37}{10}$. Now according to Art. XVII. $\frac{1}{10}$ of $\frac{475}{100}$ is $\frac{475}{1000}$, and $\frac{37}{10}$ will be 37 times as much, that is $\frac{17575}{1000} = 17\frac{575}{1000}$ $= 17.575$ as before.

In looking over the above process we find, *that the two numbers are multiplied together in the same manner as whole numbers, and as many places are pointed off for decimals in the product, as there are in the multiplicand and multiplier counted together.*

It is plain that this must always be the case, for tenths multiplied by tenths must produce tenths of tenths, that is hundredths, which is two places; tenths multiplied by hundredths must produce tenths of hundredths, or thousandths, which is three places; hundredths multiplied by hundredths must produce hundredths of hundredths, that is ten-thousandths, which is four places, &c.

What cost $5\frac{3}{4}$ tons of hay, at $27.38 per ton? $5\frac{3}{4} =$ 5.375.

$$
\begin{array}{r}
27.38 \\
5.375 \\
\hline
13690 \\
19166 \\
8214 \\
13690 \\
\hline
\$147.16750 \text{ Ans.}
\end{array}
$$

In this example there are hundredths in the multiplicand, and thousandths in the multiplier. Now hundredths multiplied by thousandths must produce hundredths of thousandths, which is five decimal places, the number found by counting the places in the multiplicand and multiplier together. The answer is 147 dollars, 16 cents, 7 mills, and $\frac{5}{10}$ of a mill.

A man owned .03 of the stock in a bank, and sold .2 of his share. What part of the whole stock did he sell?

It is evident that the answer to this question must be expressed in thousandths, for hundredths multiplied by tenths must produce thousandths. $\frac{2}{10}$ of $\frac{3}{100}$ are $\frac{6}{1000}$. But if we multiply them in the form of decimals, we obtain only one figure, viz. 6. In order to make it express $\frac{6}{1000}$ it will be necessary to write two zeros before it, thus, .006.

$$
\begin{array}{r}
.03 \\
.2 \\
\hline
\end{array}
$$

Ans. .006 of the whole stock.

This result is agreeable to the above rule.

The following is the general rule for multiplication, when there are decimals in either or both the numbers : *Multiply as in whole numbers, and point off as many places from the right of the product for decimals, as there are decimal places in the multiplicand and multiplier counted together. If the product does not contain so many places, as many zeros must be written at the left, as are necessary to make up the number.*

———◆———

Division of Decimals.

XXVIII. *A man bought* 8 *yards of broadcloth for* \$75.376 ; *how much was it per yard ?*

$$
\begin{array}{r}
\$75.376 \\
\text{mills. } 75376 \ (8 \\
72 \\
\overline{} \qquad 9422 \text{ mills.} \\
33 \\
32 \quad \$9.422 \text{ Ans.} \\
\overline{} \\
17 \\
16 \\
\overline{} \\
16 \\
16 \\
\overline{} \\
.. \\
\end{array}
$$

In this example there are decimals in the dividend only. I consider \$75.376 as 75376 mills. Then dividing by 8, either by long or short division, I obtain 9422 mills per yard, which is \$9.422. The answer has the same number of decimal places as the dividend.

Divide 117.54 *bushels of corn equally among* 18 *men. How much will each have ?*

$117.54 = 117\frac{54}{100} = {}^{1}\frac{1754}{100}$; this divided by 18 gives $\frac{1174}{18} = 6\frac{53}{100} = 6.53.$

117.54 (18
108   ——
——   6.53
95
90

'   54
    54
    ——
    . .

Or we may reason as follows.   I divide 117 by 18, which gives 6, and 9 remainder.   9 whole ones are 90 tenths, and 5 are 95 tenths ; this divided by 18 gives 5, which must be tenths, and 5 remainder.   5 tenths are 50 hundredths, and 4 are 54 hundredths ; this divided by 18 gives 3, which must be 3 hundredths.   The answer is 6.53 each, as before.

If you divide 7.75 barrels of flour equally among 13 men, how much will you give each of them ?

7.75 (13
65   ——
——   .596 +
125
117
——
80
. 78
——
2

It is evident that they cannot have so much as a barrel each. $7.75 = \frac{775}{100} = \frac{7750}{1000}$.   Dividing this by 13, I obtain $\frac{596}{1000}$ and a small remainder, which is not worth noticing, since it is only a part of a thousandth of a barrel. $\frac{596}{1000} =$ .596.   Or we may reason thus : 7 whole ones are 70 tenths, and 7 are 77 tenths.   This divided by 13 gives 5, which must be tenths, and 12 remainder.   12 tenths are 120 hundredths, and 5 are 125 hundredths.   This divided by 13 gives 9, which must be hundredths, and 8 remainder.   We may now reduce this to thousandths, by annexing a zero.   8 hundredths are 80 thousandths.   This divided by 13 gives 6, which must be thousandths, and 2 remainder.   Thousandths will be sufficiently exact in this instance, we may therefore omit the remainder. The answer is .596 + of a barrel each.

From the above examples it appears, *that when only the dividend contains decimals, division is performed as in whole numbers, and in the result as many decimal places must be pointed off from the right, as there are in the dividend.* ·

Note. If there be a remainder after all the figures have been brought down, the division may be carried further, by annexing zeros. In estimating the decimal places in the quotient, the zeros must be counted with the decimal places of the dividend.

At $6.75 a cord, how many cords of wood may be bought for $38 ?

In this example there are decimals in the divisor only. $6.75 is 675 cents or $\frac{675}{100}$ of a dollar. The 38 dollars must also be reduced to cents or hundredths. This is done by annexing two zeros. Then as many times as 675 hundredths are contained in 3800 hundredths, so many cords may be bought.

```
3800 (675      or     3800 (675
3375  ——              3375 ——
——          5 425/675 cords.   ——  5.62 + cords.
 425                  4250
                      4050
                      ————
                      2000
                      1350
                      ————
                       650
```

The answer is 5$\frac{425}{675}$ cords, or reducing the fraction to a decimal, by annexing zeros and continuing the division, 5.62 + cords.

If 3.423 yards of cloth cost $25, what is that per yard ?

$$3.423 = 3\tfrac{423}{1000} = \tfrac{3423}{1000}.$$

The question is, if $\frac{3423}{1000}$ of a yard cost $25, what is that a yard ?

According to Art. XXIV., we must multiply 25 by 1000, that is, annex three zeros, and divide by 3423.

$$
\begin{array}{ll}
25000 \ (3423 & \text{or} \\
23961 \ \text{———} \\
\text{———} \quad \$7\tfrac{1039}{3423} \\
1039
\end{array}
\qquad
\begin{array}{l}
25000 \ (3423 \\
23961 \ \text{———} \\
\text{———} \quad 7.30 + \text{Ans.} \\
10390 \\
10269 \\
\text{———} \\
121
\end{array}
$$

The answer is $\$7\tfrac{1039}{3423}$, or reducing the fraction to cents, $7.30 per yard.

If 1.875 *yard of cloth is sufficient to make a coat ; how many coats may be made of* 47.5 *yards ?*

In this example the divisor is thousandths, and the dividend tenths. If two zeros be annexed to the dividend it will be reduced to thousandths.

$$
\begin{array}{ll}
47.500 \ (1.875 & \text{or} \\
3750 \ \text{———} \\
\text{———} \quad 25\tfrac{625}{1875} \\
10000 \\
9375 \\
\text{———} \\
625
\end{array}
\qquad
\begin{array}{l}
47500 \ (1875 \\
3750 \ \text{———} \\
\text{———} \quad 25.33 + \\
10000 \\
9375 \\
\text{———} \\
6250 \\
5625 \\
\text{———} \\
6250 \\
5625 \\
\text{———} \\
625
\end{array}
$$

1875 thousandths are contained in 47500 thousandths $25\tfrac{625}{1875}$ times, or reducing the fraction to decimals, 25.33 + times, consequently, 25 coats, and $\tfrac{33}{100}$ of another coat may be made from it.

From the three last examples we derive the following rule : *When the divisor only contains decimals, or when there are more decimal places in the divisor than in the dividend, annex as many zeros to the dividend as the places in the divisor exceed those in the dividend, and then proceed as in whole numbers. The answer will be whole numbers.*

At $2.25 *per gallon, how many gallons of wine may be bought for* $15.375 *?*

'In this example the purpose is to find how many times $2.25 is contained in $15.375. There are more decimal places in the dividend than in the divisor. The first thing that suggests itself, is to reduce the divisor to the same denomination as the dividend, that is, to mills or thousandths. This is done by annexing a zero, thus, $2.250. The question is now, to find how many times 2250 mills are contained in 15375 mills. It is not important whether the point be taken away or not.

$$15375 \; (2250$$
$$13500 \; \text{———}$$
$$\text{———} \quad 6.83 + \text{gals.} \quad \text{Ans.}$$
$$18750$$
$$18000$$

$$7500$$
$$6750$$

$$750$$

Instead of reducing the divisor to mills or thousandths, we may reduce the dividend to cents or hundredths, thus, $15.375 are 1537.5 cents. The question is now, to find how many times 225 cents are contained in 1537.5 cents. This is now the same as the case where there were decimals in the dividend only, the divisor being a whole number.

$$1537.5 \; (225$$
$$1350 \; \text{———}$$
$$\text{———} \quad 6.83 + \text{gals.} \quad \text{Ans. as before.}$$
$$1875$$
$$1800$$

$$750$$
$$675$$

$$75$$

If 3.15 bushels of oats will keep a horse 1 week, how many weeks will 37.5764 bushels keep him?

The question is, to find how many times 3.15 is contained in 37.5764. The dividend contains ten thousandths. The *divisor is* 31500 ten thousandths.

```
375764 (31500
31500
───────        11.929 + weeks.   Ans.
 60764
 31500
───────
292640
283500
───────
  91400
  63000
───────
 284000
 283500
───────
    500
```

Instead of reducing the divisor to ten-thousandths, we may reduce the dividend to hundredths. 37.5764 are 3757.64 hundredths of a bushel. The decimal .64 in this, is a fraction of an hundredth.

3.15 are 315 hundredths. Now the question is, to find how many times 315 hundredths are contained in 3757.64 hundredths.

```
3757.64 (315
315
─────          11.929 + weeks.   Ans. as before.
607
315
─────
2026
2835
─────
 914
 630
─────
 2840
 2835
─────
    5
```

From the two last examples we derive the following rule for division : *When the dividend contains more decimal places than the divisor : Reduce them both to the same denomination, and divide as in whole numbers.*

N. B. There are two ways of reducing them to the same denomination. First, the divisor may be reduced to the same denomination as the dividend, by annexing zeros, and taking away the points from both. Secondly, the dividend may be reduced to the same denomination as the divisor, by taking away the point from the divisor, and removing it in the dividend towards the right as many places as there are in the divisor. The second method is preferable.

The same result may be produced by another mode of reasoning. The quotient must be such a number, that being multiplied with the divisor will reproduce the dividend. Now a product must have as many decimal places as there are in the multiplier and multiplicand both. Consequently the decimal places in the divisor and quotient together must be equal to those in the dividend. In the last example there were four decimal places in the dividend and two in the divisor ; this would give two places in the quotient. Then a zero was annexed in the course of the division, which made three places in the quotient. The rule may be expressed as follows :

Divide as in whole numbers, and in the result point off as many places for decimals as those in the dividend exceed those in the divisor. If zeros are annexed to the dividend, count them as so many decimals in the dividend. If there are not so many places in the result as are required, they must be supplied by writing zeros on the left.

Division in decimals, as well as in whole numbers, may be expressed in the form of common fractions.

What part of .5 is .3 ?   Ans. $\frac{3}{5}$.
What part of .08 is .05 ?   Ans. $\frac{5}{8}$.
What part of .19 is .43 ?   Ans. $\frac{43}{19}$.
What part of .3 is .07 ?
To answer this, .3 must be reduced to hundredths. .3 is .30, the answer therefore is $\frac{7}{30}$.
What part of 14.035 is 3.8 ?
3.8 is 3.800, the answer therefore is $\frac{3800}{14035}$.

In fine, to express the division of one number by another, when either or both contain decimals, reduce them both to the lowest denomination mentioned in either, and then write the divisor under the dividend, as if they were whole numbers.

Circulating Decimals.

XXIX. There are some common fractions which cannot be expressed exactly in decimals. If we attempt to change $\frac{1}{3}$ to decimals for example, we find .3333, &c. there is always a remainder 1, and the same figure 3 will always be repeated however far we may continue it. At each division we approximate ten times nearer to the true value, and yet we can never obtain it. $\frac{1}{6} = .1666$, &c.; this begins to repeat at the second figure. $\frac{6}{11} = .545454$, &c.; this repeats two figures. In the division the remainders are alternately 6 and 5. $\frac{56}{333} = .168168$, &c.; this repeats three figures, and the remainders are alternately 56, 227, and 272. Some do not begin to repeat until after two or three or more places. It is evident that whenever the same remainder recurs a second time, the quotient figures and the same remainders will repeat over again in the same order. In the last example for instance, the number with which we commenced was 56; we annexed a zero and divided; this gave a quotient 1, and a remainder 227; we annexed another zero, and the quotient was 6, and the remainder 272; we annexed another zero, and the quotient was 8, and the remainder 56, the number we commenced with. If we annex a zero to this, it is evident that we shall obtain the same quotient and the same remainder as at first, and that it will continue to repeat the same three figures for ever.

It is evident that the number of these remainders, and consequently the number of figures which repeat, must be one less than the number of units in the divisor. If the fraction is $\frac{1}{7}$, there can be only six different remainders; after this number, one of them must necessarily recur again, and then the figures will be repeated again in the same order.

```
1  (7
10 —
7 .1428571, &c.
—
30
28
—
20
14
—
60
56
—
·40
35
/
50
49
—
10
7
—
3
```

It commences with 1 for the dividend, then annexing zeros, the remainders are 3, 2, 6, 4, 5, which are all the numbers below 7; then comes 1 again, the number with which it commenced, and it is evident the whole will be repeated again in the same order. Decimals which repeat in this way are called *circulating deci mals.*

Whenever we find that a fraction begins to repeat, we may write out as many places as we wish to retain, without the trouble of dividing.

As it is impossible to express the value of such a fraction by a decimal exactly, rules have been invented by which operations may be performed on them, with nearly as much accuracy as if they could be expressed; but as they are long and tedious, and seldom used, I shall not notice them. Sufficient accuracy may always be attained without them.

I shall show, however, how the true value of them may always be found in common fractions.

The fraction $\frac{1}{9}$ reduced to a decimal, is .1111 ... &c. Therefore, if we wish to change this fraction to a common fraction, instead of calling it $\frac{1}{10}$, $\frac{11}{100}$, or $\frac{111}{1000}$, which will be a value too small, whatever number of figures we take, we must call it $\frac{1}{9}$. This is exact, because it is the fraction which produces the decimal. If we have the fraction .2222.. &c. it is plain that this is twice as much the other, and must *be called* $\frac{2}{9}$. If $\frac{2}{9}$ be reduced to a decimal, it produces .2222 ... &c. If we have .3333 ∴ &c. this being three times as much as the first, is $\frac{2}{6} = \frac{1}{3}$. If $\frac{1}{3}$ be reduced to a decimal, it produces .3333 . . &c. It is plain, that whenever a single figure repeats, it is so many ninths.

Change .4444 &c. to a common fraction. Ans. $\frac{4}{9}$.

Change .5555 &c. to a common fraction.

Change .6666 &c. to a common fraction.

Change .7777 &c. to a common fraction.

Change .9999 &c. to a common fraction.

Change .5333 &c. to a common fraction.

This begins to repeat at the second figure or hundredths. The first figure 5 is $\frac{5}{10}$; and the remaining part of the fraction is $\frac{3}{9}$ of $\frac{1}{10}$, that is, $\frac{3}{90} = \frac{1}{30}$; these must be added together. $\frac{5}{10}$ is $\frac{15}{30}$, and $\frac{1}{30}$ makes $\frac{16}{30} = \frac{8}{15}$. The answer is $\frac{8}{15}$. If this be changed to a decimal, it will be found to be .5333 &c.

If a decimal begins to repeat at the third place, the two first figures will be so many hundredths, and the repeating figure will be so many ninths of another hundredth.

Change .4666 &c. to a common fraction.

Change .3888 &c. to a common fraction.

Change .3744 &c. to a common fraction.

Change .46355 &c. to a common fraction.

If $\frac{1}{99}$ be changed to a decimal, it produces .010101 &c. The decimal .030303 &c. is three times as much, therefore it must be $\frac{3}{99} = \frac{1}{33}$. The decimal .363636 &c. is thirty-six times as much, therefore it must be $\frac{36}{99} = \frac{4}{11}$.

If $\frac{1}{999}$ be changed to a decimal, it produces .001001001 &c. The decimal .006006 &c. is 6 times as much, therefore it must be $\frac{6}{999} = \frac{2}{333}$. The fraction .027027 &c. is twenty-seven times as much, and must be $\frac{27}{999} = \frac{3}{111}$. The fraction .354354 &c. is 354 times as much, and must be $\frac{354}{999} = \frac{118}{333}$. This principle is true for any number of places. Hence we derive the following rule for changing a circulating decimal to a common fraction: *Make the repeating figures the numerator, and the denominator will be as many 9s as there are repeating figures.*

If they do not begin to repeat at the first place, the preceding figures must be called so many tenths, hundredths, &c. according to their number, then the repeating part must be changed in the above manner, but instead of being the fraction of an unit, it will be the fraction of a tenth, hundredth, &c. according to the place in which it commences.

Instead of writing the repeating figures over several times,

18 *

they are sometimes written with a point over the first and last to show which figures repeat. Thus .333 &c. is written .3̇. .2525 &c. is written .2̇5̇. .387387 &c. is written .3̇8̇7̇. .57346346 &c. is written .5̇7346̇. -

Change .2̇4̇ to a common fraction.

Change .4̇2̇ to a common fraction.

Change .5̇3̇7̇ to a common fraction.

Change .4̇745̇ to a common fraction.

Change .8̇374̇ to a common fraction.

Change .4̇7647̇ to a common fraction.

Note. To know whether you have found the right answer, change the common fraction, which you have found, to a decimal again. If it produces the same, it is right.

Proof of Multiplication and Division by casting out 9s.

If either the multiplicand or the multiplier be divisible by 9, it is evident the product must be so.
Multiply 437 by 85.

437		81 times 437 = 35397
85		4 times 432 = 1728
		4 times 5 = 20
2185		
3496		37145

Ans. 37145

85 = 81 + 4, and 437 = 432 + 5. 81 is divisible by 9, and 85 being divided by 9 leaves a remainder 4. 432 is divisible by 9, and 437 leaves a remainder 5. 81 times 437, and 4 times 432, and 4 times 5, added together, are equal to 85 times 437. 81 times 437 is divisible by 9, because 81 is so, and 4 times 432 is divisible by 9, because 432 is so. The only part of the product which is not divisible by 9, is the product of the two remainders 4 and 5. This product, 20, divided by 9, leaves a remainder 2. It is plain, therefore, that if the whole product, 37145, be divided by 9, the remainder must be 2, the same as that of the product of the remainder.

Therefore to prove multiplication, *divide the divisor and the dividend by* 9, *and multiply the remainders together, and divide the product by* 9, *and note the remainder ; then divide the whole product by* 9, *and if the remainder is the same as the last, the work is right.*

Instead of dividing by 9, the figures of each number may be added, and their sum be divided by 9, as in Art. XXI., (and for the same reason) and the remainders will be the same as if the numbers themselves were divided.

In the above example, say 7 and 3 and 4 are 14, which, divided by 9, leaves a remainder 5; then 5 and 8 are 13, which, divided by 9, leaves a remainder 4. Then 4 times 5 are 20, which, divided by 9, leaves a remainder 2. Then adding the figures of the product, 5 and 4 and 1 and 7 and 3 are 20, which being divided by 9 leaves 2, as the other. Instead of dividing 14 and 13 by 9, these figures may be added together, thus 4 and 1 are 5; 3 and 1 are 4.

Since in division the quotient multiplied by the divisor produces the dividend ; *if the divisor and quotient be divided by* 9 *and the remainders multiplied together, and this product divided by* 9, *and the remainder noted; and then the dividend be divided by* 9 ; *this last remainder must agree with the other.*

N. B. If there is a remainder after division, it must be subtracted from the dividend before proving it.

Miscellaneous Examples.

1. If 2 lbs. of figs cost 2s. 8d., what is that per lb. ?

2. If 2 bushels of corn cost 8s. 6d., what is that per bushel ?

3. If 2 lbs. of raisins cost 1s. 10d., what is that per lb. ?

4. If 3 bushels of potatoes cost 9s. 6d., what is that per bushel ?

5. If 4 gals. of gin cost 12s. 8d., what is that per gal. ?

6. If 2 barrels of flour cost 3£. 4s., what is that per barrel ?

7. If 2 gallons of wine cost 1£. 10s. 4d., what is that per gallon ?

8. If 2 barrels of beer cost 1£. 15s. 8d., what is that per barrel ?

9. If 4 gallons of gin cost 17s. 8d., what is that per gallon?

10. Ir 5 yards of cloth cost 6£. 10s. 5d., what is that per yard ?

11. If 7 barrels of flour cost 17£. 8s. 7d., what is that per barrel ?

12. If 8 yards of cloth cost 20£. 18s. 5., what is that per yard ?

13. A man had 4 cwt. 3 qrs. 14 lbs. of tobacco, which he put into 2 boxes, $\frac{1}{2}$ of it in each ; how much did he put in each box ?

14. Divide 13£. 8s. 5d. equally among 5 men.

15. Divide 8 cwt. 3 qrs. 17 lbs. into 3 equal parts.

16. Divide 16 cwt. 1 qr. 11 lbs. of flour equally among 7 men ; how much will each have ?

17. Divide 3 hhds. 42 gals. 2 qts. into 5 equal parts.

18. If 12 yards, 3 qrs. 2 nls. of cloth will make 7 coats, how much will make 1 coat ? How much will make 13 coats ?

19. If 5 yards of cloth cost 19£. 3s. 4d., what cost 17 yards ?

20. What is $\frac{2}{5}$ of 45£. 9s. 7d. ?

21. If 18 cwt. of sugar cost 56£. 13s. 8d. what will 53$\frac{2}{5}$ cwt. cost ?

22. If $\frac{4}{7}$ of a ship is worth 943£. 7s. 8d., what is the whole ship worth ?

23. If 84 cows cost 453£. 14s. 8d., how much is that apiece ?

24. If 3$\frac{1}{4}$ cwt. of sugar cost 9£. 15s. 9d., what is that per cwt. ?

25. If 9$\frac{3}{4}$ barrels of flour cost 21£. 3s. 8d., what cost 17$\frac{1}{2}$ barrels ?

26. If a staff 4 feet long cast a shade on level ground 6 ft. 8 in., what is the height of a steeple which casts a shade 173 feet at the same time ?

27. If 57 gallons of water in one hour run into a cistern allons, and by another cock 42 gallons ru

afternoon of the same day, and travels at the rate of $6\frac{1}{4}$ miles per hour. The distance is 250 miles. Supposing them to travel constantly until they meet, at what time will they meet, and at what distance from each place ?

30. The distance from New-York to Baltimore is 197 miles. Two travellers set out at the same time in order to meet; A from New-York towards Baltimore, and B from Baltimore towards New-York. When they met, which was at the end of 6 days, A had travelled 3 miles a day more than B. How many miles did each travel per day ?

31. If when wheat is 7s. 6d. per bushel, the penny-loaf weighs 9 oz., what ought it to weigh when wheat is 6s. per bushel ?

32. Suppose 650 men are in a garrison, and have provisions sufficient to last them two months; how many men must leave the garrison in order to have the provisions last those who remain five months ?

33. If 8 boarders will drink a barrel of cider in 15 days, how long will it last if 4 more boarders come among them ?

34. A ship's crew of 18 men is supposed to have provision sufficient to last the voyage, if each man is allowed 23 oz. per day, when they pick up a crew of 8 persons. What must then be the daily allowance of each person ?

35. How many yards of flannel that is $1\frac{1}{4}$ yard wide will line a cloak, containing 9 yards, that is $\frac{4}{5}$ yard wide ?

36. A garrison of 1800 men have provisions sufficient to last them 12 months; but at the end of 3 months, the garrison was reinforced by 600 men, and 2 months after that, a second reinforcement of 400 men was sent to the garrison. How long did the provisions last in the whole ?

37. A regiment of soldiers, consisting of 1000, are to be new clothed; each coat to contain $2\frac{1}{2}$ yards of cloth $1\frac{1}{4}$ yard wide, and to be lined with flannel of $\frac{3}{4}$ yard wide. How many yards of flannel will line them ?

38. I borrowed 185 quarters of corn, when the price was 19s. per quarter; how much must I pay to indemnify the lender when the price is 17s. 4d. ?

39. If 7 men can reap 84 acres of wheat in 12 days, how many men can reap 100 acres in 5 days ?

40. If 7 men can build 36 rods of wall in 3 days, how many rods can 20 men build in 14 days ?

41. If 20 bushels of wheat are sufficient for a family of 15

persons 3 months, how much will be sufficient for 4 persons 11 months ?

42. If it cost $23.84 to carry 17 cwt. 3 qrs. 14 lb. 85 miles, how much must be paid for carrying 53 cwt. 2 qrs. 150 miles ?

43. If 18 men can build a wall 40 rods long, 5 feet high, and 4 feet thick in 15 days ; in what time will 20 men build one 87 rods long, 8 feet high, and 5 feet thick ?

44. If a family of 9 persons spend $305 in 4 months, how many dollars would maintain them 8 months, if 5 persons more were added to the family ?

45. If a regiment consisting of 1878 soldiers, consume 702 quarters of wheat in 336 days ; how many quarters will an army of 22536 soldiers consume in 112 days ?

46. If 12 tailors can finish 13 suits of clothes in 7 days, how many tailors can finish the clothes of a regiment consisting of 494 soldiers, in 19 days of the same length ?

47. If 24 measures of wine, at 3s. 4d. serve 16 men for 6 days, how many measures, at 2s. 8d., will serve 48 men 4 days ?

48. How many tiles 8 inches square, will cover a hearth 12 feet wide and 16 feet long ?

49. How many bricks 9 in. long, 4½ in. wide, and 2 in. thick, will build a wall 6 feet high and 13½ in. thick, round a garden, each side of which is 280 feet on the outside of the wall ?

50. There is a house 40 feet in length, and 30 feet rafters ; how many shingles will it take to cover the roof, supposing each shingle to be 4 inches wide, and each course to be 6 inches ?

51. A man built a house consisting of 4 stories ; in the lower story there were 16 windows, each containing 12 panes of glass, each pane 16 in. long, 12 in. wide ; the second and third stories contained 18 windows, each of the same size; the fourth story contained 18 windows, each window 6 panes 18 by 12. How many square feet of glass were there in the whole house ?

52. A merchant sold a piece of cloth for $40, and by so doing lost 10 per cent. He ought in trading to have gained 15 per cent. For how much ought he to have sold the cloth ?

53. *Bought a hogshead of molasses for $25, but 12 gallons having leaked out, I desire to sell the remainder, so as to*

gain 3 per cent. on the whole cost. For how much per gallon must I sell it ? ϸ *ʘ, S⁻*

54. Bought a hogshead of brandy, for $93 on 6 months' credit, and sold it for $103 ready money. How much did I gain, allowing money to be worth 6 per cent. a year ? *⸫ ᑐ, ᶣ Ƴ*

55. Bought 3 hhds. of wine for $320 ready money, and sold it at $1.87 per gal. on 6 months' credit. What did I gain, allowing money to be worth 6 per cent. per year ? *⸫ ᶥ, ᶣ ᶥ*

Note. To answer this question, it will be necessary to compute the interest on $320 for 6 months, and add it to $320.

56. Bought a quantity of goods for $437.45 and hired the money to pay for it, for which I paid at the rate of 8 per cent. a year. Having kept it on hand 3 months and 17 days, I sold it for $470, on 4 months' credit. What per cent. did I gain ?

57. Bought 5 hhds. of rum at 1 dollar per gal., ready money, and having kept it 3 months and 23 days, I sold it at $1.20 per gallon, on 5 months' credit ; 16 gals. had leaked out while in my possession. How much did I gain ?

When a debtor keeps money longer than a year, the interest is considered as due to the creditor at the end of the year, and he has a right to demand it. If the interest is not paid at the end of the year, the creditor sometimes requires the interest for the year to be added to the principal, and considered a part of the debt, and consequently interest paid upon it for the rest of the time, and so on at the end of every year. In this way the principal increases every year by the interest of the last year. This may seem just, but it is not allowed by law. This is called *compound interest.*

58. What will $143.17 amount to in 3 years and 4 months, at 6 per cent. compound interest ?

The most convenient method is, to find the amount of 1 dollar for the time, and then multiply it by the number of dollars in the question.

$20,

ample have
st ?
.53 for 11 years,

May, 1813, for $847, rate 6
The following payments were

1.00
.06

.06 interest for 1 year.
+ 1.00

= 1.06 amount for 1 year.
.06

.0636 interest for 2d year.
+ 1.06

= 1.1236 amount for 2 years.
.06

.06744 interest for 3d year.
+ 1.1236

= 1.191016 amount for 3 years.
.02   rate for 4 months.

.02382032 interest for 4 months.
+ 1.191016

= 1.21483632 amount for 3 years and 4 months.

It will be sufficiently exact to use the first four **decimals** $1.2148. This multiplied by 143.17 will give the answer.

1.2148
143.17

85036
12148
36444
48592
12148

$173.922916   Ans. $173.923—.

52. A make a table which shall contain the amount of 1 dol-
doing lost 10 par, for two years, for 3 years, &c. to 20 years,
15 per cent. tt. and at 6 per cent. Reserve five decimal
cloth ?

53. Bought a hogshead on will serve for sterling money, or
lons having leaked out, I desire ßressed in decimals.

years	5	rates	6	years	5	rates	6
1	1.05000	1.06000		11			
2	1.10250	1.12360		12			
3				13			
4				14			
5				15			
6				16			
7				17			
8				18			
9				19			
10				20			

60. What is the compound interest of $17.25 for 2 years and 7 months, at 5 per cent. ?

Note. From the table take the amount of 1 dollar for two years, at 5 per cent. and compute the interest on it for .7 months, at 5 per cent. as in simple interest ; add this to the amount for two years. This will be the amount of 1 dollar for 2 years and 7 months. Multiply this by 17.25 ; this will be the amount of $17.25 for the time. Then to find the interest, subtract the principal from the amount.

61. What will $73.42 amount to in 4 years, 3 months, and 17 days, at 6 per cent. compound interest ?

62. A note was given 13th March, 1815, for $847.25 ; how much had it amounted to on the 7th November, 1820, at 6 per cent. compound interest ?

63. How much would the sum in the last example have amounted to in the same time at simple interest ?

64. What is the compound interest of $1753 for 11 years, 10 months, and 22 days, at 6 per cent. ?

65. A note was given 11th May, 1813, for $847, rate 6 per cent. compound interest. The following payments were

19

made; 18th February, 1815, $158; 19th December, 1816, $87; 5th October, 1819, $200. What was due 8th July, 1822?

66. What will 17£. 13s. 6d. amount to in 5 years, 3 months, at 6 per cent. compound interest?

Note. Change the shillings and pence to decimals of a pound, and proceed as in Federal money. Call the unit in the table 1£. instead of 1 dollar.

67. What is the compound interest of $643, for 7 years, 5 months, and 18 days, at 5 per cent.?

68. What is the compound interest of 143£. 7s. 4d. for 19 years, 7 months, at five per cent.?

69. A farmer mixed 15 bushels of rye, at 64 cents per bushel; 18 bushels of corn, at 55 cents per bushel; and 21 bushels of oats, at 28 cents per bushel. How many bushels were there of the mixture? What was the whole worth? What was it worth per bushel?

70. A grocer mixed 123 lb. of sugar, that was worth 8 cents per lb.; 87 lb. that was worth 11 cents per lb.; and 15 lb. that was worth 13 cents per lb. What was the mixture worth per lb.?

71. A grocer mixed 43 gallons of wine, that was worth $1.25 per gal. with 87 gals. that was worth $1.60 per gal. What was the mixture worth per gal.?

72. With a hhd. of rum, worth $.87 per gal. a grocer mixed 10 gals. of water. What was the mixture worth per gal.?

73. How many gals. of rum, at $.60 per gal. will come to · as much as 43 gals. will come to, at $.75 per gal.?

74. How much water must be added to a pipe of wine, worth $1.50 per gal. in order to reduce the price to $1.30 per gal.?

75. A grocer has two kinds of sugar, one at 8 cents per lb., the other at 13 cents. He wishes to mix them together in such a manner, that the mixture may be worth 11 cents per lb. What will be the proportions of each in the mixture?

Note. The difference of the two kinds is 5 cents. Therefore if a pound of each kind be divided, each into five equal parts, the difference between one part of each will be 1 cent. If ⅘ lb. be taken from that at 8 cents, and ⅕ lb. of that at 13 cents be put in its place, the pound will be worth 9 cents. If ⅖ lb. be taken from it, and as much of the other be put in its place, the pound will be worth 11 cents, as required. The pound then will consist of $\frac{2}{5}$, at 8 cents, and $\frac{3}{5}$, at 13 cents. If 5 lb. be mixed, there will be 2 lb. at 8, and 3 at 13 cents. The proportions are 2 lb. at 8 as often as 3 lb. at 13 cents.

76. A farmer had oats, at 38 cents per bushel, which he wished to mix with corn, at 75 cents per bushel, so that the mixture might be 50 cents per bushel. What were the proportions of the mixture?

Note. The difference in the price of a bushel is 37 cents. The difference between $\frac{1}{37}$ of a bushel of each is 1 cent. If $\frac{12}{37}$ of a bushel be taken from a bushel of oats, and $\frac{12}{37}$ of a bushel of corn be put in its place, a bushel will be formed worth 50 cents, and consisting of $\frac{12}{37}$ corn, and $\frac{24}{37}$ oats. The proportions are 12 of oats to 25 of corn.

It is easy to see that *the denominator will always be the difference of the prices of the ingredients, and the difference between the mean and the less price will be the numerator for the quantity of the greater, and the difference between the mean and the greater will be the numerator for the quantity of the less value. Take away the denominators, and the numerators will express the proportions.*

77. A merchant has spices, some at 9d. per lb. some at 1s., some at 2s. and some at 2s. 6d. per lb. How much of each sort must he mix, that he may sell the mixture at 1s. 8d. per lb.?

Note. Take one kind, the price of which is greater, and one, the price of which is less than the mean, and find the proportions as above. Then take the other two and find their proportions in the same way.

Less 9d. = 9d.⎫ mean ⎧11d. diff. between less
⎬ 20 ⎨ and mean.
Greater 2s. 6d. = 30d.⎭ ⎩10d. diff. between great er and mean.

The proportions are 10 of the less to 11 of the greater.

Less 1s. = 12d.⎫ mean ⎧8d. diff. between less
⎬ 20 ⎨ and mean.
Greater 2s. = 24d.⎭ ⎩4d. diff. between great- er and mean.

The proportions are 4 of the less to 8 of the greater, which is the same as 1 of the less to 2 of the greater.

The answer is 10 lb. at 9d. to 11 lb. at 2s. 6d., and 1 lb. at 1s. to 2 lb. at 2s.

Other proportions might be found by comparing the first, and third, and the second and fourth.

78. A grocer has two sorts of tea, one at 75 cents per lb. and the other at $1.10 per lb. How must he mix them in order to afford the mixture at $1.00 per lb. ?

79. A grocer would mix the following kinds of sugar, viz. at 10 cents, 13 cents, and 16 cents per lb. What quantity of each must he take to make a mixture worth 12 cents per lb. ?

Note. Those at 13 and 16 must both be compared with that at 10 cents separately. .

80. A grocer has rum worth $.75 per gal. ; how many parts water must he put in, that he may afford to sell the mixture at $.65 per gal. ?

81. It is required to mix several sorts of rum, at 5s. 7d., and 9s. per gal. with water, so that the mixture may be worth 6s. per gal. How much of each sort must the mixture consist of ?

82. A farmer had 10 bushels of wheat, worth 8s. per bushel, which he wished to mix with corn, at 3s. per bushel, so that the mixture might be worth 5s. per bushel. How many bushels of corn must he use ?

Note. Find the proportions for a single bushel as before, then find how much corn must be put with 1 bushel of wheat, and then with 10 bushels. The proportions are 2 of wheat to 3 of corn, consequently 1 of wheat to 1½ of corn, and 10 of wheat to 15 of corn.

83. A farmer would mix 20 bushels of rye, at 65 cents per bushel, with barley at 51 cents, and oats at 30 cents per bushel. How much barley and oats must be mixed with rye, that the mixture may be worth 41 cents per bushel ?

84. A grocer had 43 gallons of wine worth $1.75 per gal., which he wished to mix with another kind worth $1.40 per gal., so that the mixture might be worth $1.60 per gal. How many gals. of the latter kind must he use ?

85. Three merchants, A, B, and C, freight a ship with wine. A put on board 500 tons, B 340, and C 94 ; in a

storm they were obliged to cast 150 tons overboard. What loss does each sustain ?

See Part 1. Art. XVI., example 158 and following.

86. A father dying, bequeathed an estate of $12000 as follows : ⅓ to his wife, ⅓ to his eldest son, ¼ to his second son, and ⅙ to his daughter. It is required to divide the estate in these proportions.

Note. Reduce the fractions to a common denominator, and the numerators will show the proportions.

87. Two men hired a pasture for $37, A put in 3 horses for 4 months, and B 5 horses for 3 months. What ought each to pay ?

Note. 3 horses for 4 months is the same as 4 times 3 or 12 horses for 1 month ; and 5 horses for 3 months, is the same as 3 times 5, or 15 horses for 1 month. The question therefore is the same, as if A had put in 12 horses and B 15. A must pay $\frac{12}{27}$ and B $\frac{15}{27}$, or, reducing the fractions, $\frac{4}{9}$ and $\frac{5}{9}$.

88. Two men, A and B, traded in company : A put in $350 for 8 months, and B $640 for 5 months ; they gained $250. What was the share of each ?

Note. Make the time equal, as in the last example.

89. Four men jointly hired a pasture for 20 English guineas ; A turned in 7 oxen for 13 days, B 9 oxen for 14 days, C 11 oxen for 25 days, and D 15 oxen for 37 days. How much ought each to pay ?

90. A family of 10 persons took a large house for ⅓ of a year, for which they were to pay $500, for that time. At the end of 14 weeks they took in 4 new lodgers ; and after 3 weeks, 4 more ; and so on for every 3 weeks, during the term, they took in 4 more lodgers. What must one of each class pay per week of the rent ? -

91. Three men enter into partnership and trade as follows : A put in 150£., and at the end of 7 months took out 50£. ; 5 months after he put in 170£ ;—B put in 205£., and at the end of 5 months, 110£. more, but took out 150£. 4 months after ;—C put in 300 guineas, at 28s. each, and when 8 months had elapsed, he drew out 150£., but 5 months after he put in 500£. Their partnership continued 18 months, at the end of which time they had gained 450£. *Required* each person's share of the gain.

19 *

92. The last five are examples of *compound* or *double fellowship*. What rule can you make for it?

93. In how long time will 1 dollar gain as much interest as $15 will gain in 1 month?

94. In how long time will 1 dollar gain as much interest as 8 dollars will gain in 3 months?

95. In how long time will 1 dollar gain as much interest as 24 dollars will gain in 5 months?

96. In how long time will 1 dollar gain as much interest as $158 will gain in 11 months?

97. In how long time will 3 dollars gain as much interest as 1 dollar will gain in 24 months?

98. In how long time will 28 dollars gain as much interest as 1 dollar will gain in 157 months?

99. A lent B 8 dollars for 2 months, afterwards B lent A 1 dollar; how long ought he to keep it to satisfy him for the former favour?

100. C lent D 1 dollar for 15 months; afterwards D lent C 5 dollars; how long ought he to keep it to satisfy him for the former favour?

101. A borrowed of B 17 dollars for 11 months, promising him a like kindness; afterwards B lent A 25 dollars. How long ought he to keep it?

Note. Find how long he ought to keep 1 dollar, and then how long he ought to keep 25 dollars.

102. I lent a friend 257 dollars, which he kept 15 months, promising to do me a like kindness, but he was not able to let me have more than 100 dollars; how long ought I to keep it?

103. A owes B notes to be paid as follows: 7 dollars to be paid in 3 months, and 5 dollars to be paid in 8 months; but he wishes to pay the whole at once. In what time ought he to pay it?

Note. 7 dollars for 3 months is the same as 1 dollar for 21 months; and 5 dollars for 8 months is the same as 1 dollar for 40 months. $40 + 21 = 61$, and $7 + 5 = 12$. He might have 1 dollar 61 months; the question now is how long he may keep 12 dollars. It is evident he might keep it $\frac{1}{12}$ of 61 months.

104. C owes D $380, to be paid as follows; $100 in 6 months; $120 in 7 months; and $160 in 10 months. He

wishes to pay the whole at once. In how long a time ought he to pay it ?

105. A merchant has due to him 300£. to be paid as follows ; 50£. in 2 months ; 100£ in 5 months ; and the rest in 8 months. It is agreed to make one payment of the whole. In what time ought he to receive it ?

106. F owes H $1000, of which $200 is to be paid present, $400 in 5 months, and the rest in 15 months. They agree to make one payment of the whole. Required the time ?

107. A merchant has due a certain sum of money, of which ⅛ is to be paid in 2 months, ⅓ in 3 months, and the rest in 6 months. In what time ought he to receive the whole ?

108. A merchant has three notes due to him as follows : one of $300 due in 2 months ; one of $250 due in 5 months; and one of $180 due 3 months ago ; the whole of which he wishes to receive now. What ought he to receive, allowing 6 per cent. interest ?

Note. First find the equated time, and then the interest or discount for present payment, as shall be found necessary.

$300 for 2 months = 1 dol. for 600 months.
$250 for 5 months = 1 dol. for 1250 months.

—————

1850

The two notes not yet due are the same as 1 dollar for 1850 months. But he has had $180 3 months after it was due, which is the same as 1 dollar for 540 months. This must be taken out of the other, and there will remain 1 dollar for 1310 months. If he can have 1 dollar for 1310 months, how long can he have $730 ?

131,0 (73,0
73   —————
—   1.8 nearly = 1 **month and 24 days.**
580
584

As it is not due until 1 month and 24 days after this time, it must be discounted for that time. See Part 1. Art. XXIV., example 130 and following. 6 per cent for 1 year is ₉⁄₁₀ per cent. or .009 for 1 month and 24 days. The fraction then is ₁₀₀₀/₁₀₀₀. $730 is ₁₀₀₀/₁₀₀₀ of what ?

109. A gave B four notes as follows ; one of $75, dated 5th June, 1819, to be paid in 4 months; one of $150, dated 15th August, to be paid in 6 months ; one of $170, dated 11th September, to be paid in 5 months ; and one of $300 dated 15th November, to be paid in 3 months. They were all without interest until they were due. On 1st January, 1820, he proposed to pay the whole. What ought he to pay ?

110. A owes B $158.33, due in 11 months and 17 days, without interest, which he proposes to pay at present. What ought he to pay, when the rate of money is 5 per cent. ?

Note. The rate per cent. for 11 mo. 17 days, at 5 per cent. a year, is about $4\frac{8}{10}$ per cent. or .048, consequently the amount of 1 doll. is $1.048. $158.33 is $\frac{15833}{1000}$ of the num ber.

It is easy to find the rate per cent. of the discount for any given time, when the rate of interest is given. When interest is 6 per cent., that is, $\frac{6}{100}$, the discount is $\frac{6}{106}$, because the discount of 106 dolls. is 6 dolls. If $\frac{6}{106}$ be converted into a decimal, it gives the rate of discount in decimals, so that it may be computed in the same manner as interest. This changed to a decimal is .0566. .057 — is sufficiently exact. This is $5\frac{7}{10}$ per cent. The rate must be found for the time required, before it is changed to a decimal.

In the last example the fraction would be $\frac{48}{1048}$, which is .046 nearly. Multiply the sum by this, and you will have the discount, which subtracted from the sum, will be the answer required.

111. What is the discount of $143.87 for 1 year and 5 months, when interest is 6 per cent. ?

112. What is the present worth of a note of $84.67, due in 1 year, 3 months, and 14 days, without interest, when the rate of interest is $5\frac{1}{2}$ per cent. ?

113. A man has a note of $647 due in 2 years and 7 months, without interest ; but being in want of the money, he sells the note ; what ought he to receive, when the usual rate of interest is 6 per cent. ?

114. A gentleman divided $50 between two men, A and B. A's share was $\frac{3}{7}$ of B's. What was the share of each ?

Note. This question is to divide the number 50 into two parts, that shall be in the proportion of 3 and 7 ; that is, one

shall have 3 as often as the other shall have 7.　$7 + 3 =$.
10.　A had $\frac{3}{10}$ and B $\frac{7}{10}$.

115. A gentleman bequeathed an estate of $12500 between his wife and son.　The son's share was $\frac{7}{}$ of the share of the wife.　What was the share of each ?

116. What is the hour of the day, when the time past from midnight is equal to $\frac{5}{11}$ of the time to noon ?

117. Two men talking of their ages, one says $\frac{2}{3}$ of my age is equal to $\frac{3}{}$ of yours : and the sum of our ages is 95. What were their ages ?

Note. To find the proportions, reduce them to a common denominator and take the numerators.

118. If a man can do $\frac{3}{8}$ of a piece of work in one day, in what part of a day can he do $\frac{1}{8}$ of it ?　How long will it take him to do the whole ?

119. A farmer hired two men to mow a field ; one of them could mow $\frac{1}{3}$ of it in a day, and the other $\frac{1}{5}$ of it.　What part of it would they both together do in a day ?　How long would it take them both to mow it ?

120. A gentleman hired 3 men to build a wall ; the first could do it alone in 8 days, the second in 10 days, and the third in 12 days.　What part of it could each do in a day ? How long would it take them all together to finish it ?

121. A man and his wife found that when they were together, a bushel of corn would last 15 days, but when the man was absent, it would last the woman alone 27 days. What part of it did both together consume in 1 day ?　What part did the woman alone consume ?　What part did the man alone consume ?　How long would it last the man alone ?

122. Three men lived together, one of them found he could drink a barrel of cider alone in 4 weeks, the second could drink it alone in 6 weeks, and the third in 7 weeks. How long would it last the three together ?

123. A cistern has 3 cocks to fill it, and one to empty it. One cock will fill it alone in 3 hours, the second in 5 hours, and the third in 9 hours. ·The other will empty it in 7 hours.　If all the cocks are allowed to run together, in what time will it be filled ?

124. Divide 25 apples between two persons, so as to give one 7 more than the other.

Note. Give one of them 7, and then divide the rest equally.

125. A gentleman divided an estate of $15000 between his two sons, giving the elder $2500 more than the younger. What was the share of each ?

126. A gentleman bequeathed an estate of $50000, to his wife, son, and daughter; to his wife he gave $1500 more than to the son, and to the son $3500 more than to the daughter. What was the share of each ?

127. A, B, and C, built a house, which cost $35000; A paid $500 more, and C $300 less than B. What did each pay ?

128. A man bought a sheep, a cow, and an ox, for $62 ; for the cow he gave $10 more than for the sheep; and for the ox $10 more than for both. What did he give for each ?

129. A man sold some calves and some sheep for $108 ; the calves at $5, and the sheep at $8 apiece. There were twice as many calves as sheep. What was the number of each sort ?

Note. There were two calves and one sheep for every $18.

130. A farmer drove to market some oxen, some cows, and some sheep, which he sold for $749 ; the oxen at $28, the cows at $17, and the sheep at $7.50. There were twice as many cows as oxen, and three times as many sheep as cows. How many were there of each sort ?

131. A man sold 16 bushels of rye, and 12 bushels of wheat for £8. 16s. The wheat at 3s. per bushel more than the rye. What was each per bushel ?

Note. The whole of the wheat came to 36s. more than the same number of bushels of rye. Take out 36s., and the remainder will be the price of 28 bushels of rye.

132. Four men, A, B, C, and D, bought an ox for $50, which they agreed to share as follows : A and B were to have the hind quarters, C and D the fore quarters. The hind quarters were considered worth $\frac{1}{4}$ cent per lb. more than the fore quarters. A's quarter weighed 217 lb.; B's 223 lb. ; C's 214 lb. ; and D's 219 lb. The tallow weighed 73 lb., which they sold at 8 cents per lb. ; and the hide 43 lb., which they sold at 5 cents per lb. What ought each to pay ?

133. At the time they bought the above ox, the fore quarters of beef were worth 6 cents per lb., and the hind quarters 6½ cents per lb. It is required to find what each ought to pay in this proportion.

Note. This is a more just manner of dividing the cost, than that in the last example. It may be done by finding what the quarters would come to, at this rate, and then dividing the real cost in that proportion.

134. Said A to B, my horse and saddle together are worth $150, but my horse is worth 9 times as much as the saddle. What was the value of each?

135. A man driving some sheep and some cattle, being asked how many he had of each sort, said he had 174 in the whole, and there were $\frac{9}{20}$ as many cattle as sheep. Required the number of each sort.

136. A man driving some sheep, and some cows, and some oxen, being asked how many he had of each sort, answered, that he had twice as many sheep as cows, and three times as many cows as oxen; and that the whole number was 80. Required the number of each sort.

137. A gentleman left an estate of $13000 to his four sons, in such a manner, that the third was to have once and one half as much as the fourth, the second was to have as much as the third and fourth, and the first was to have as much as the other three. What was the share of each?

138. A, B, and C playing at cards, staked 324 crowns; but disputing about the tricks, each man took as many crowns as he could get. A got a certain number; B as many as A, and 15 more; and C ¼ part of both their sums added together. How many did each get?

139. The stock of a cotton manufactory is divided into 32 shares, and owned equally by 8 persons, A, B, C, &c. A sells 3 of his shares to a ninth person, who thus becomes a member of the company, and B sells 2 of his shares to the company, who pay for them from the public stock. After this, A wishes to dispose of the remainder of his part. What proportion of the whole stock does he own?

140. Three persons, A, B, and C, traded in company. A put in $75; B $40; and C a sum unknown. They gained $64, of which C took $18 for his share. What did C put in?

141. *How many cubic feet in a cistern, 4 ft. 2 in. long, 3 ft. 8 in. wide, and 2 ft. 7 in. high?*

A method of doing this by decimals has already been shown. It is now proposed to do it by a method called *duodecimals*.

First, I find the square feet in the bottom of the cistern.

$$3 \text{ ft. } 8 \text{ in.} = 3\tfrac{8}{12} \text{ ft.} \qquad 4 \text{ ft. } 2 \text{ in.} = 4\tfrac{2}{12} \text{ ft.}$$

$$4\tfrac{2}{12}$$
$$3\tfrac{8}{12}$$

$$2 \text{ ft. } 7 \text{ in.} = 2\tfrac{7}{12} \text{ ft.}$$

$$2\tfrac{9}{12} + \tfrac{4}{144}$$
$$12\tfrac{6}{12}$$

$$\overline{\qquad\qquad}$$

$$15\tfrac{3}{12} + \tfrac{4}{144} \text{ square feet in the bottom.}$$
$$2\tfrac{7}{12}$$

$$\overline{\qquad\qquad}$$

$$8\tfrac{10}{12} + \tfrac{11}{144} + \tfrac{4}{1728}$$
$$30\tfrac{6}{12} + \tfrac{8}{144}$$

Ans. $39\tfrac{5}{12} + \tfrac{7}{144} + \tfrac{4}{1728}$ cubic feet in the cistern.

I say $\tfrac{8}{12}$ of $\tfrac{2}{12}$ is $\tfrac{16}{144} = \tfrac{1}{12} + \tfrac{4}{144}$, I write down the $\tfrac{4}{144}$ and reserve the $\tfrac{1}{12}$; then $\tfrac{8}{12}$ of 4 is $\tfrac{32}{12}$ and $\tfrac{1}{12}$ (which was reserved) is $\tfrac{33}{12} = 2\tfrac{9}{12}$, which I write down. Then 3 times $\tfrac{2}{12}$ is $\tfrac{6}{12}$, and 3 times 4 are 12. These added together make $15\tfrac{3}{12} + \tfrac{4}{144}$ square feet. Then to find the cubic feet, I multiply this by $2\tfrac{7}{12}$. $\tfrac{7}{12}$ of $\tfrac{4}{144}$ is $\tfrac{28}{1728} = \tfrac{2}{144} + \tfrac{4}{1728}$, I write the $\tfrac{4}{1728}$, and reserve the $\tfrac{2}{144}$; then $\tfrac{7}{12}$ of $\tfrac{3}{12}$ is $\tfrac{21}{144}$, and $\tfrac{2}{144}$ (which were reserved) are $\tfrac{23}{144} = \tfrac{1}{12} + \tfrac{11}{144}$; I write down the $\tfrac{11}{144}$ and reserve the $\tfrac{1}{12}$; then $\tfrac{7}{12}$ of 15 are $8\tfrac{9}{12}$ and $\tfrac{1}{12}$ (which was reserved) is $8\tfrac{10}{12}$. 2 times $\tfrac{4}{144}$ are $\tfrac{8}{144}$; and 2 times $\tfrac{3}{12}$ are $\tfrac{6}{12}$, and 2 times 15 are 30. Adding them together, $\tfrac{8}{144}$ and $\tfrac{11}{144}$ are $\tfrac{19}{144} = \tfrac{1}{12} + \tfrac{7}{144}$; I write the $\tfrac{7}{144}$, and reserve the $\tfrac{1}{12}$; then $\tfrac{10}{12}$ and $\tfrac{6}{12}$ are $\tfrac{16}{12}$, and $\tfrac{1}{12}$ (which was reserved) is $\tfrac{17}{12} = 1\tfrac{5}{12}$. The whole is $39\tfrac{5}{12} + \tfrac{7}{144} + \tfrac{4}{1728}$.

Since we know that 12ths multiplied by 12ths will produce 144ths, and that $\tfrac{12}{144}$ make $\tfrac{1}{12}$; and, also, that 144ths multiplied by 12ths produce 1728ths, and that $\tfrac{12}{1728}$ make $\tfrac{1}{144}$, we may write the fractions without their denominators, if we make some mark to distinguish one from the other. It is usual to distinguish 12ths by an accent, thus ('), 144ths thus (''), 1728ths thus ('''), &c. 12ths are called primes; *144ths* seconds; 1728ths thirds, &c.

Operation.

```
 4   2′
 3   8′
 ─────────
 2   9′   4″
12   6′
 ─────────
15   3′   4″
 2   7′
 ─────────
 8  10′  11″  4‴
30   6′   8″
 ─────────
```

Cubic feet 39   5′   7″   4‴

The operation is precisely the same as before. To adopt the language suited to this notation, *we say, units multiplied by primes or primes by units produce primes, seconds by units produce seconds, &c. primes by primes produce seconds, seconds by primes produce thirds. Also 12 thirds make 1 second, 12 seconds 1 prime, 12 primes make 1 foot, whether long, square, or cubic. The same principle extends to fourths, fifths, &c.*

142. How much wood in a load 4 ft. 8 in. high, 3 ft. 11 in. broad, and 8 ft. long ?

Note.. Multiply the height and breadth together, and divide by 2. See page 102.

143. How many square feet in a floor 16 ft. 8 in. wide, and 18 ft. 5 in. long ?

144. How much wood in a pile 4 ft. wide, 3 ft. 8 in. high, and 23 ft. 7 in. long ?

145. If 11 barrels of cider will buy 4 barrels of flour, and 7 barrels of flour will buy 40 barrels of apples ; what will 1 barrel of apples be worth, when cider is $2.50 per barrel ?

146. A person buys 12 apples and 6 pears for 17 cents, and afterwards 3 apples and 12 pears for 20 cents. What is the price of an apple and of a pear ?

Note. At the second time he bought 3 apples and 12 pears for 20 cents, 4 times all this will make 12 apples and 48 pears for 80 cents ; the price of 12 apples and 6 pears being taken from this, will leave 63 cents for 42 pears, which is 1½ cent apiece.

147. Two persons talking of their ages, one says $\frac{2}{3}$ of mine is equal to $\frac{3}{4}$ of yours, and the difference of our ages is 10 years. What were their ages ?

148. A gentleman divided some money among 4 persons, giving the first as much as the second and fourth ; the second as much as the third and fourth ; the third, half as much as the first ; and the fourth, 5 cents. How much did he give to each ?

149. Two persons, A and B, talking of their ages, A says to B, $\frac{1}{6}$ of mine and $\frac{1}{4}$ of yours are equal to 13 ; B says to A, $\frac{1}{2}$ of mine and $\frac{1}{8}$ of yours are equal to 16. What was the age of each ?

150. A person drew two prizes ; $\frac{1}{4}$ of the first, and $\frac{1}{3}$ of the second was \$120 ; and the sum of the two was \$400. What was each prize ?

151. Two persons purchase a house for \$4200 ; the first could pay for the whole, if the second would give him $\frac{1}{3}$ of his money ; and the second could pay for the whole, if the first would give him $\frac{1}{5}$ of his money. How much money had each.

152. A man bought some lemons at 2 cents each, and $\frac{3}{4}$ as many, at 3 cents each, and then sold them all at the rate of 5 cents for 2, and by so doing gained 25 cents. How many lemons did he buy ?

153. There are two cisterns which receive the same quantity of water ; the first constantly loses $\frac{1}{6}$ of what it receives ; after running 7 days, 10 barrels were taken from the second, and then the quantity of water in the two was equal. How much water did each receive per day ?

154. A man having \$100 spent a certain part of it ; he afterwards received five times as much as he spent, and then his money was double what it was at first. How much did he spend ?

155. A man left his estate to 2 sons and 3 daughters, each son had 5 dollars as often as each daughter had 4 ; the difference between the sum of the sons' shares and that of the daughters, was \$1000. Required the share of a son.

156. A man left his estate to his wife, son, and daughter, as follows : to his wife $\frac{1}{3}$ of the whole, and $\frac{1}{3}$ as much as the share of the daughter ; to his son $\frac{1}{3}$ of the whole, and to the daughter the remainder, which was \$1000 less than the share of the son. What was the share of each ?

157. A man bought some oranges for 25 cents ; if he had

bought 3 less for the same money, the price of an orange would have been once and a half of the price he gave. What was the price of an orange ?

158. A man divided his estate among his children as follows : to the first he gave twice as much as to the third, and to the second two thirds as much as to the first ; the portion of the second and third together was $1500. What was the portion of each ?

159. A man bought 16 bushels of corn, and 20 bushels of rye for $30 ; and also 24 bushels of corn, and 10 of rye for $27. How much per bushel did he give for each ?

160. A man travelling from Boston to Philadelphia, a distance of 335 miles, at the expiration of 7 days, found that the distance which he had to travel was equal to $\frac{2}{4}\frac{4}{5}$ of the distance which he had already travelled. How many miles per day did he travel ?

161. A man left his estate to his three sons ? the first had $2000, the second had as much as the first, and $\frac{1}{3}$ as much as the third, and the third as much as the other two. What was the share of each ?

162. A man when he married was three times as old as his wife ; $\frac{}{}$5 years afterwards he was but twice as old. What was the age of each when they were married ?

163. A grocer bought a cask of brandy, $\frac{1}{4}$ of which leaked out, and he sold the remainder, at $1.80 per gal., and by that means received for it as much as he gave. How much did it cost him per gal. ?

164. A and B laid out equal sums of money in trade ; A gained a sum equal to $\frac{1}{4}$ of his stock, and B lost $2,25 ; then A's money was double that of B. What did each lay out ?

165. There is a fish whose head is 16 inches long, his tail is as long as his head and half the length of his body, and his body is as long as his head and tail. What is the length of the fish ?

166 There are three persons, A, B, and C, whose ages are as follows : A is 20 years old, B is as old as A and $\frac{4}{7}$ of the age of C, and C is as old as A and B both. What are the ages of B and C ?

167. A person has two silver cups and only one cover. The first cup weighs 12 oz. If the first cup be covered, it will weigh twice as much as the second, but if the second cup be covered, it will weigh three times as much as the first. Required the weight of the cover and of the second cup.

168. Three persons do a piece of work; the first and second together do $\frac{2}{3}$ of it, and the second and third together do $\frac{7}{11}$. What part of it is done by the second?

169. A man bought apples, at 5 cents per doz., half of which he exchanged for pears, at the rate of 8 apples for 5 pears; he then sold all his apples and pears, at 1 cent each, and by so doing gained 19 cents. How many apples did he buy, and how much did they cost?

170. A man being asked the hour of the day, answered that it was between 7 and 8, but a more exact answer being required, said the hour and minute hands were exactly together. Required the time.

171. What is the hour of the day when the time past from noon is equal to $\frac{5}{17}$ of the time to midnight?

172. What is the hour of the day when $\frac{3}{7}$ of the time past from midnight is equal to $\frac{1}{5}$ of the time to noon?

173. A merchant laid out $50 for linen and cotton cloth, buying 3 yards of linen for a dollar, and 5 yards of cotton for a dollar. He afterwards sold $\frac{1}{4}$ of his linen, and $\frac{1}{5}$ of his cotton for $12, which was 60 cents more than it cost him. How many yards of each did he buy?

174. A gentleman divided his fortune among his three sons, giving A 8 as often as B 5, and B 7 as often as C 4; the difference between the shares of A and C was $7500. What was the share of each?

175. A tradesman increased his estate annually by $150 more than the fourth part of it; at the end of 3 years it amounted to 14811\frac{7}{18}$. What was it at first?

176. A hare has 50 leaps before a grey-hound, and takes 4 leaps to his 3; but two of the grey-hound's leaps are equal to 3 of the hare's. How many leaps must the grey-hound take to overtake the hare?

177. A labourer was hired for 60 days, upon this condition, that for every day he worked he should receive $1.50; and for every day he was idle, he should forfeit $.50; at the expiration of the time he received $75. How many days did he work?

178. A and B have the same income, A saves $\frac{1}{5}$ of his, but B, by spending 30£. a year more than A, at the end of 8 years finds himself 40£. in debt. What is their income, and what does each spend per year?

179. A lion of bronze, placed upon the basin of a fountain, can spout water into the basin through his throat, his eyes, and his right foot. If he spouts through his throat only, he will fill the basin in 6 hours; if through his right eye only, he will fill it in 2 days; if through his left eye only, he will fill it in 3 days; if through his right foot only, he will fill it in 4 hours. In what time will the basin be filled if the water flow through all the apertures at once?

180. A player commenced play with a certain sum of money; at the first game he doubled his money, at the second he lost 10 shillings, at the next game he doubled what he then had, at the fourth game he lost 20 shillings; twice the sum he then had was as much less than 200s., as three times the sum would be greater than 200s. Required the sum with which he commenced play.

181. What is the circumference of a wheel of which the diameter is 5 feet?

The circumference of a circle is 3.1416, or more exactly 3.1415926 times the diameter.

182. What is the diameter of a wheel of which the circumference is 17 feet?

A *parallelogram* is a figure with four sides in which the opposite sides are parallel or equidistant throughout their whole extent. In the adjacent figure A B C D is a parallelogram, and also A B E F. A B E F is a rectangular parallelogram, or a rectangle, and is measured as explained page 79. It is easy to see that A B C D is equal to A B E F, because the triangle B C E is equal to A D F. The contents of a parallelogram, then, is found by multiplying the length of one of its sides as A B, by the perpendicular which measures the distance from that side to its opposite, as B E.

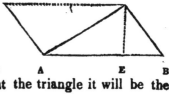

The triangle A is half the parallelogram A B C D. The area of a triangle, therefore, will be half the product of the base A B, by the perpendicular C E. If the perpendicular should fall without the triangle it will be the same.

To find the area of any irregular figure, divide it into triangles.

20 *

To find the area of a circle, multiply half the diameter by half the circumference. Or multiply half the diameter into itself, and then multiply it by 3.1415926.

To find the solid contents of a round stick of timber, find the area of one end, and multiply it by the length.

If a round or a square stick tapers to a point, it contains just ⅓ as much as if it were all the way of the same size as at the largest end. If the stick tapers but does not come to a point, it is easy to find when it would come to a point, and what it would then contain, and then to find the contents of the part supposed to be added, and take it away from the whole.

183. What is the area of a parallelogram, of which one side is 13 feet, and the perpendicular 7 feet?

Ans. 91 *square feet.*

184. How much land is in a triangular field, of which one side is 28 rods, and the distance from the angle opposite that side to that side, 15 rods?

Ans. 210 *sq. rods, or* 1 *acre and* 50 *rods.*

185. How many square inches in a circle, the diameter 10 inches? *Ans.* 78.54 + *in.*

186. How many solid feet in a round stick of timber 10 inches in diameter and 17 feet long?

Ans. 9.272 + *ft.*

187. How many cubic feet of water will a round cistern hold which is 3 ft. in diameter at the bottom, 4 ft. at top, and 5 ft. high? *Ans.* 48.433 *ft.*

———◆———

Geographical and Astronomical Questions.

188. The diameter of the earth is 7911.73 miles; what is its circumference?

189. The earth turns round once in 24 hours; how far are the inhabitants at the equator carried each hour by this motion?

190. The circumference of the earth is divided into 360 degrees; how many miles in a degree?

191. How many degrees does the earth turn in 1 hour?

192. How many minutes of a degree does the earth turn *in 1 minute* of time?

193. What is the difference in the time of two places whose difference of longitude is 23° 43' ?

194. The longitude of Boston is 71° 4' W. of Greenwich, England. What is the time at Greenwich when it is 11 h. 43 min. morn. at Boston ?

195. The long. of Philadelphia is 75° 09' W., that of Rome 12° 29' E. What is the time at Philadelphia, when at Rome it is 6 h. 27 min. even. ?

196. The earth moves round the sun in 1 year, in an orbit nearly circular. Its distance from the sun is about 95,000,000 of miles ; what distance does the earth move every hour ?

197. The lat. of Turk's Island is 21° 30' N. and the long. is about the same as that of Boston. The lat. of Boston is 42° 23' N. How many miles apart are they ?

198. The mouth of the Columbia river is about 125° W. long., and Montreal is about 73¼ W. long., they are in about the same lat. A degree of longitude in that latitude is about 48.3 miles. How many miles are they apart, measuring on a parallel of latitude ?

Examples in Exchange.

It is not necessary to give rules for exchange. There are books which explain the relative value of foreign and American coin, weights, and measures. The one may be exchanged to the other by multiplication or division.

199. What is the value of 13£. 14s. 8d. English or sterling money, in Federal money ?

It will be most convenient to reduce the shillings and pence to the decimal of a pound. For the value, see the table.

200. What is the value of $153.78 in sterling money ?

201. What is the value of 853 francs, 50 centimes, in Federal money ?

202. What is the value of $287.42, in French money ?

203. What is the value of 523 Dutch gelders or florins, at 40 cents each, in Federal money ?

204. What is the value of $98.59 in Dutch gelders.

205. What is the value of 387 ducats of Naples, at $777¼ each, in Federal money ?

Tables of Coin, Weights, and Measures.

Denominations of Federal money as determined by an **Act** of Congress, Aug. 8, 1786.

10 mills make	1 cent	marked c.
10 cents	1 dime	d.
10 dimes	1 dollar	$
10 dollars	1 Eagle	E.

The coins of Federal money are two of gold, four of silver, and two of copper. The gold coins are an eagle and half-eagle; the silver, a dollar, half-dollar, double-dime, and dime; the copper, a cent and half-cent. The standard gold and silver is eleven parts fine, and one part alloy. The weight of fine gold in the eagle is 246.268 grains; of fine silver in the dollar, 375.64 grains; of copper in 100 cents 2¼ lbs. avoirdupois.*

ENGLISH MONEY.

4 farthings make	1 penny d. value in U. S.	$0.019
12 pence	1 shilling s.	.228
20 shillings	1 pound £.	4.4444
21 shillings	1 guinea	4.6724

FRENCH MONEY.

100 centimes make 1 franc, value $.1875.

TROY WEIGHT.

24 grains (gr.) make	1 penny-weight	dwt.
20 dwt.	1 ounce	oz.
12 oz.	1 pound	lb.

By this weight are weighed jewels, gold, silver, corn, bread, and liquors.

APOTHECARIES' WEIGHT.

20 grains (gr.) make	1 scruple	sc.
3 sc.	1 dram	dr. or ℨ
8 dr.	1 ounce	oz. or ℥
12 oz.	1 lb.	

* The above are the coins which were at first contemplated, but the double-dime has never been coined. Twenty-five-cent pieces and half-dimes have been coined.

Apothecaries use this weight in compounding their medicines; but they buy and sell their drugs by Avoirdupois
weight. Apothecaries' is the same as Troy, having only some
different divisions.

AVOIRDUPOIS WEIGHT.

16 drams (dr.) make	1 ounce	oz.
16 oz.	1 pound	lb.
28 lbs.	1 quarter	qr.
4 qrs.	1 hundred-weight	cwt.
20 cwt.	1 ton	T.

By this weight are weighed all things of a coarse and
drossy nature; such as butter, cheese, flesh, grocery wares,
and all metals except gold and silver.

DRY MEASURE.

2 pints (pt.) make	1 quart	qt.
8 qts.	1 peck	pk.
4 pks.	1 bushel	bu.
8 bu.	1 quarter	qr.

The diameter of a Winchester bushel is $18\frac{1}{2}$ inches, and
its depth 8 inches.—And one gallon by dry measure contains $268\frac{4}{5}$ cubic inches.

By this measure salt, lead ore, oysters, corn, and other dry
goods are measured.

ALE OR BEER MEASURE.

2 pints (pt.) make	1 quart	qt.
4 qts.	1 gallon	gal.
8 gals.	1 firkin of ale	fir.
9 gals.	1 firkin of beer	fir.
2 fir.	1 kilderkin	kil.
2 kil.	1 barrel	bar.
3 kil.	1 hogshead	hhd.
3 bar.	1 butt	butt.

The ale gallon contains 282 cubic inches. In London
the ale firkin contains 8 gallons, and the beer firkin 9; other
measures being in the same proportion.

WINE MEASURE.

2 pints (pt.) make	1 quart	qt.
4 qts	1 gallon	gal.
42 gals.	1 tierce	tier.
63 gals.	1 hogshead	hhd.
84 gals.	1 puncheon	pun.
2 hhds.	1 pipe or butt	p. or b.
2 pipes	1 tun	T.
18 gals.	1 runlet	run.
$31\frac{1}{2}$ gallons	1 barrel	bar.

The wine gallon contains 231 cubic inches.

By this measure brandy, spirits, perry, cider, mead, vinegar, and oil are measured.

CLOTH MEASURE.

$2\frac{1}{4}$ inches make	1 nail	nl.
4 nls.	1 quarter	qr.
4 qrs.	1 yard	yd.
3 qrs.	1 ell Flemish	Ell Fl.
5 qrs.	1 ell English	Ell Eng.
5 qrs.	1 aune or ell French.	

The French aune is 42 inches.

LONG MEASURE.

3 barley corns make	1 inch	in.
12 in.	1 foot	ft.
3 ft.	1 yard	yd.
$5\frac{1}{2}$ yds.	1 pole or rod	pole
40 poles	1 furlong	fur.
8 fur.	1 mile	ml.
3 mls.	1 league	l.
60 geographical miles, or		
$69\frac{1}{2}$ statute miles	1 degree nearly,	deg. or °
360 degrees the circumference of the earth.		

Also, 4 inches make	1 hand
5 feet	1 geometrical pace
6 feet	1 fathom
6 points	1 line
12 lines	1 inch

SQUARE MEASURE.

144	inches make 1 foot		ft.
9	ft	1 yard	yd.
30¼ yds. or 272¼ ft.	}	1 pole, rod, or perch.	
40	poles	1 rood.	
4	roods	1 acre.	

CUBIC OR SOLID MEASURE.

1728	inches make	1 foot	ft.
27	feet	1 yard.	
40 feet of round timber, or 50 feet of hewn timber	}	1 ton or load.	
128	solid feet	1 cord of wood.	

TIME.

60	seconds make	1 minute	m.
60	minutes	1 hour	h.
24	hours	1 day	d.
7	days	1 week	w.
4	weeks	1 month	
13 months, 1 day, and 6 hours or 365 days, 6 hours	}	1 Julian year Y	
12	calendar months	1 year.	

The true length of the solar year is 365 days, 5 hours, 48 min. 57 seconds.

Reflections on Mathematical Reasoning.

If the learner has studied ' 1e preceding pages attentively, he has had some practice in mathematical reasoning. It may now be pleasant, as well as useful, to give some attention to the principles of it.

By attending to the objects around us, we observe two properties by which they are capable of being increased or diminished, viz. in number and extent.

Whatever is susceptible of increase and diminution is the object of mathematics.

Arithmetic is the science of numbers.

All individual or single things are naturally subjects of number. Extent of all kinds is also made a subject of number, though at first view it would seem to have no connexion with it. But to apply number to extent, it is necessary to have recourse to artificial units. If we wish to compare two distances, we cannot form any correct idea of their relative extent, until we fix upon some length with which we are familiar as a measure. This measure we call *one* or a *unit*. We then compare the lengths, by finding how many times this measure is contained in them. By this means length becomes an object of number. We use different units for different purposes. For some we use the inch, for others the foot, the yard, the rod, the mile, &c.

In the same manner we have artificial units for surfaces, for solids, for liquids, for weights, for time, &c. And in all there are different units for different purposes.

When a measure is assumed as a unit, all smaller measures are fractions of it. If the foot is taken for the unit, inches are fractions. If the rod is the unit, yards, feet, and inches are fractions, and the smaller, being fractions of the larger, are fractions of fractions. It may be remarked, that all parts are properly units of a lower order. As we say single things are units, so when they are cut into parts, these parts are single things, and consequently units, and they are numbered as such. When a thing is divided into eight equal parts, for example, the parts are numbered, one, two, three, &c. As we put together several units and make a collec tion which is called a unit of a higher order, so any single thing may be considered as a collection of parts, and these parts will be units of a lower order. The unit may be considered as a collection of tenths, the tenths as a collection of hundredths, &c.

The first knowledge we have of numbers and their uses is derived from external objects; and in all their practical uses they are applied to external objects. In this form they are called *concrete numbers*. Three horses, five feet, seven dollars, &c. are *concrete* numbers.

When we become familiar with numbers, we are able to think of them and reason upon them without reference to any particular object, as three, five, seven, four times three are twelve, &c. These are called *abstract numbers*.

Though all arithmetic operations are actually performed on abstract numbers, yet it is generally much easier to reason upon concrete numbers, because a reference to sensible objects shows at once the purpose to be obtained, and at the same time, suggests the means to arrive at it, and shows also how the result is to be interpreted.

Success in reasoning depends very much upon the perfections of the language which is applied to the subject, and also upon the choice of the words which are to be used. The choice of words again depends chiefly on the knowledge of their true import. There is no subject on which the language is so perfect as that of mathematics. Yet even in this there is great danger of being led into errors and difficulties, for want of a perfect knowledge of the import of its terms. There is not much danger in reasoning on concrete numbers; but in abstract numbers persons pretty well skilled in mathematics, are sometimes led into a perfect paradox, and cannot discover the cause of it, when perhaps a single word would remove the whole difficulty. This usually happens in reasoning from general principles, or in deriving particular consequences from them. The reason is, the general principles are but partially understood. This is to be attributed chiefly to the manner in which mathematics are treated in most elementary books, where one general principle is built upon another, without bringing into view the particulars on which they are actually founded.

There are several different forms in which subtraction may appear, as may be seen by referring to Art. VIII. In order to employ the word subtraction in general reasoning, either of the operations ought readily to bring this word to mind, and the word ought to suggest either of the operations.

The word division would naturally suggest but one purpose, that is, to divide a number into parts ; but it is applied to another purpose, which apparently has no immediate connexion with it, viz. to discover how many times one number is contained in another. In fractions the terms multiplication and division are applied to operations, which neither of the terms would naturally suggest. The process of multiplying a whole number by a fraction (Art. XVI.) is so different from what is called multiplication of whole numbers, that it requires a course of reasoning to show the connexion, and much practice, to render the term familiar to this operation. These remarks apply to many other instances, but they apply with much greater force to the division of whole numbers by fractions. Arts. XXIII. and XXIV. are instances of this. It is difficult to conceive that either of these, and more especially the latter, is any thing like division ; and it is still more difficult to conceive that the operations in these two articles come under the same name. When a person learns division of whole numbers by fractions from general principles, where neither of these operations is brought into view, it is easy to conceive how very imperfect his idea of it will be. The truth is, (and I have seen numerous instances of it,) that if he happens to meet with a practical case like those in the articles mentioned above, any other term in the world would be as likely to occur to him as division. In an abstract example the difficulty would be very much increased.

The above observations suggest one practical result, which will apply to mathematics generally, and it will be found to apply with equal force to every other subject. In adopting any general term or expression, we should be careful to examine it in as many ways as possible. Secondly, we should be careful not to use it in any sense in which we have not examined it. Thirdly, if we find any difficulty in using it in a case where we are sure it ought to apply, it is *an indication* that we do not fully understand it in that sense, and that it requires further examination.

I shall give a few instances of errors and difficulties into which persons, not sufficiently acquainted with the principles, sometimes fall.

Suppose a person has obtained a knowledge of the rule of division by a course of abstract reasoning, and that the only definite idea that he attaches to it is, that it is the opposite of multiplication, or that it is used to divide a number into parts. Let him pursue his arithmetic in this way, and learn to divide a whole number by a fraction. He will be astonished to find a quotient larger than the dividend; and if the divisor be a decimal, his astonishment will be still greater, because the reason is not so obvious. Let him divide 40 by $\frac{4}{9}$ according to the rule, and he will find a quotient 90. Or let him divide 45 by .03 and he will find a quotient 1500. This seems a perfect paradox, and he will be quite unable to account for it. Now if he had the idea intimately joined with the term division, that the quotient shows how many times the divisor is contained in the dividend ; and also a proper idea of a fraction, that it is less than *one*, instead of saying, divide 40 by $\frac{4}{9}$, or 45 by .03, he would say, how many times is $\frac{4}{9}$ contained in 40, or .03 in 45 ; and all the difficulty would vanish.

Innumerable instances occur, which show the importance of a single idea attached to a general term, which the term itself would not readily bring to mind, but which a single word is often sufficient to recal. The most important accessory ideas to be attached to the term division are, that the quotient shows how many times the divisor is *contained* in the dividend ; and that it is the reverse of multiplication. Those for subtraction are that it shows the *difference* of the two numbers ; and that it is the reverse of addition.

Sometimes, it is asked if dollars and pounds, or gallons be multiplied together, what will they produce ? If dollars be divided by dollars, what will they produce ? If dollars be divided by bushels, what will they produce ? &c.

It is observed, in square measure, that the length multiplied by the breadth gives the number of square feet in any rectangular surface. It is sometimes asked, if dollars be multiplied by dollars, what will be produced ? If 5s. 3d. be multiplied by 3s. 8d., what will be the result ?

It is observed in fractions, that tenths divided by tenths, hundredths by hundredths, &c. produce units ; from this some have concluded, that a cent divided by a cent, or a mill by a mill, would produce a dollar, and though they are aware of the absurdity, cannot tell how to avoid the conclusion.

The above difficulties arise chiefly from not making a proper distinction between abstract and concrete numbers. Not one of these cases can ever occur in the manner here proposed. They are imperfect examples. When a perfect example is proposed, which involves one of the above cases, the difficulty is entirely removed.

It is not proper to speak of dollars being multiplied or divided by dollars or gallons.

At 5 dollars per barrel, what cost 3 barrels of flour ?

Instead of saying that 5 dollars is to be multiplied by 3 barrels, say 3 barrels will cost three times as much as 1 barrel, that is three times 5 dollars.

If 1 dollar will buy 7 lbs. of raisins, how many pounds may be bought for 4 dollars ?

Say 4 dollars will buy 4 times as many pounds as 1 dollar. In these two examples there is no doubt what the answer should be. In one it is dollars, and in the other it is pounds.

In a piece of cloth 5 feet long and 3 feet wide, how many square feet ?

If it were 5 feet long and 1 foot wide, it would contain 5 square feet, but being 3 feet wide it will contain three times as many, or three times 5 feet.

In a certain town a tax was laid of 1 dollar upon every $150 ; how much did a man possess whose tax was 3 dollars ?

It is evident that he possessed three times $150.

At 1 cent each, how many apples may be bought for 1 cent ?

Here the divisor is 1 cent and the dividend is 1 cent, and the result is an apple instead of a dollar.

How many gallons of wine at 2 dollars per gal., may be bought for 6 dollars ?

As many times as 2 dollars are contained in 6 dollars, so many gallons may be bought.

The truth is, the numbers are always used as abstract numbers, but a reference to particular objects is kept in view, and the nature of the question will always show to what the result must be applied.

It may however be established as a general principle, that

the multiplier and multiplicand are never applied to the same object, and in precisely the same way ; and the product will be applied to the object which is mentioned in one denomination, as being the value of a unit in the other.

In division there are two numbers given to find a third, two of which will always be of the same denomination, and the other different, or differently applied.

If the divisor and dividend are of the same denomination and applied in the same way, the question is, to find how many times the one is contained in the other, and the quotient will be applied differently.

If the divisor and the dividend are of different denominations, or differently applied to the same denomination, the question is to divide the dividend into parts, and the quotient will be applied in the same manner as the dividend.

When any difficulty occurs in solving a question, it is best to supply very small numbers, and solve it first with them, and then with the numbers given. If the question is in an abstract form, endeavour to form a practical one, which shall require the same operation, and the difficulty is generally very much diminished.

In all cases reason from many to one, or from a part to one ; and then from one to many or to a part. If several parts be given, always reason from them to one part, and then to many parts, or to the whole.

IMPROVED

SCHOOL BOOKS.

Colburn's First Lessons, or, Intellectual Arithmetic.

THE merits of this little work are so well known, and so highly appreciated in Boston and its vicinity, that any recommendation of it is unnecessary, except to those parents and teachers in the country, to whom it has not been introduced. To such it may be interesting and important to be informed, that the system of which this work gives the elementary principles, is founded on this simple maxim ; that, *children should be instructed in every science, just so fast as they can understand it.* In conformity with this principle, the book commences with examples so simple, that they can be perfectly comprehended and performed mentally by children of four or five years of age ; having performed these, the scholar will be enabled to answer the more difficult questions which follow. He will find, at every stage of his progress, that what he has already done has perfectly prepared him for what is at present required. This will encourage him to proceed, and will afford him a satisfaction in his study, which can never be enjoyed while performing the merely mechanical operation of ciphering according to artificial rules.

This method entirely supersedes the necessity of any rules, and the book contains none. The scholar learns to reason correctly respecting all combinations of numbers ; and if he reasons correctly, he must obtain the *desired* result. The scholar who can be made to understand how a sum *should* be done, needs neither book nor instructer to dictate how it *must* be done.

This admirable elementary Arithmetic introduces the scholar at once to that simple, practical system, which accords with the natural operations of the human mind. All that is learned in this way is precisely what will be found essential in transacting the ordinary business of life, and it prepares the way, in the best possible manner, for the more abstruse investigations which be- long to maturer age. Children of five or six years of age will be able to make considerable progress in the science of numbers by pursuing this simple method of studying it; and it will uniformly be found that this is one of the most useful and interesting sciences upon which their minds can be occupied. By using this work children may be farther advanced at the age of nine or ten, than they can be at the age of fourteen or fifteen by the common method. Those who have used it, and are regarded as competent judges, have uniformly decided that more can be learned from it in one year, than can be acquired in two years from any other treatise ever published in America. Those who regard economy in time and money, cannot fail of holding a work in high estimation which will afford these important advantages.

Colburn's First Lessons are accompanied with such instructions as to the proper mode of using them, as will relieve parents and teachers from any embarrassment. The sale of the work has been so extensive, that the publishers have been enabled so to reduce its price, that it is, at once, the cheapest and the best Arithmetic in the country.

Colburn's Sequel.

This work consists of two parts, in the first of which the author has given a great variety of questions, arranged according to the method pursued in the First Lessons ; the second part consists of a few questions, with the solution of them, and such copious illustrations of the principles involved in the examples in the first part of the work, that the whole is rendered perfectly intelligible. The two parts are designed to be studied together. The answers to the questions in the first part are given in a Key, which is published separately for the use of instructers. If the scholar find any sum difficult, he must turn to the principles and illustrations, given in the second part, and these will furnish all the assistance that is needed.

The design of this arrangement is to make the scho lar understand his subject thoroughly, instead of per forming his sums by rule.

The First Lessons contain only examples of numbers so small, that they can be solved without the use of a slate. The Sequel commences with small and simple combinations, and proceeds gradually to the more exten sive and varied, and the scholar will rarely have occasion for a principle in arithmetic, which is not fully illustrated in this work.

Colburn's Introduction to Algebra.

THOSE who are competent to decide on the merits of this work, consider it equal, at least, to either of the others composed by the same author.

The publishers cannot desire that it should have a higher commendation. The science of Algebra is so much simplified, that children may proceed with ease and advantage to the study of it, as soon as they have finished the preceding treatises on arithmetic. The same method is pursued in this as in the author's other works ; every thing is made plain as he proceeds with his subject.

The uses which are performed by this science, give it a *high* claim to more general attention. Few of the more abstract mathematical investigations can be conducted without it; and a great proportion of those, for which arithmetic is used, would be performed with much greater facility and accuracy by an algebraic process.

The study of Algebra is singularly adapted to discipline the mind, and give it direct and simple modes of reasoning, and it is universally regarded as one of the most pleasing studies in which the mind can be engaged.

The Author's Preface.

The first object of the author of the following treatise has been to make the transition from arithmetic to algebra as gradual as possible. The book, therefore, commences with practical questions in simple equations, such as the learner might readily solve without the aid of algebra. This requires the explanation of only the signs plus and minus, the mode of expressing multiplication and division, and the sign of equality; together with the use of a letter to express the unknown quantity. These may be understood by any one who has a tolerable knowledge of arithmetic. All of them, except the use of the letter, have been explained in arithmetic. To reduce such an equation, requires only the application of the ordinary rules of arithmetic; and these are applied so simply, that scarcely any one can mistake them, if left entirely to himself. One or two questions are solved first with little explanation in order to give the learner an idea of what is wanted, and he is then left to solve several by himself.

The most simple combinations are given first, then those which are more difficult. The learner is expected to derive most of his knowledge by solving the examples himself; therefore care has been taken to make the explanations as few and as brief as is consistent with giving an idea of what is required.

In order to study this work to advantage, the learner should solve every question in course, and do it algebra-

cally. If he finds a question which he can solve as easily without the aid of algebra as with it, he may be assured, this is what the author expected. If he first solves a question, which involves no difficulty, he will understand perfectly what he is about, and he will thereby be enabled to encounter those which are difficult.

When the learner is directed to turn back and do in a new way, something he has done before, let him not fail to do it, for it will be necessary to his future progress; and it will be much better to trace the new principle in what he has done before than to have a new example for it.

The author has heard it objected to his arithmetics by some, that they are too easy. Perhaps the same objection will be made to this treatise on algebra. But in both cases, if they *are* too easy, it is the fault of the subject, and not of the book. For in the First Lessons, there is no explanation; and in the Sequel there is probably less than in any other books, which explain at all. As easy however as they are, the author believes that whoever undertakes to teach them, will find the intellects of his scholars more exercised in studying them, than in studying the most difficult treatise he can put into their hands.

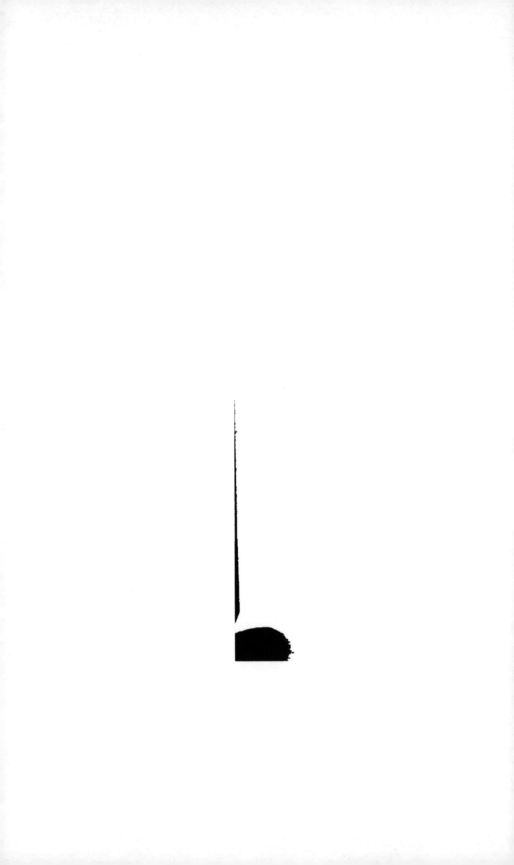

CATALOGUE

OF

School & Classical Books.

HILLIARD, GRAY, & Co.

BOSTON,

ARE EXTENSIVELY ENGAGED IN PUBLISHING A GREAT VARIETY OF

THE MOST VALUABLE

ELEMENTARY BOOKS.

☞ IT is their aim to publish only such as possess superior merits ; and that every work shall be well printed on good paper, and well bound. Several of the School Books, which they have heretofore published, have been revised and re-modelled, and are now presented to the public in an improved form. They have also an Extensive Assortment of Valuable Works in all the Departments of Literature and Science. They solicit the patronage of the Public, and invite Instructers and Literary Gentlemen to examine these Books.

☞ *Liberal Discounts made to School Committees and wholesale purchasers.*

The AMERICAN FIRST CLASS BOOK, or Exercises in Reading and Recitations, selected principally from Modern Authors of Great Britain and America, and designed for the use of the highest Class in public and private Schools. By JOHN PIERPONT, Minister of Hollis street Church, Boston. Author of Airs of Palestine, &c. Price $1,00.

Extract from the Preface.

This Book has been compiled with a special reference to the Public Reading and Grammar Schools of this City. It is the result of an attempt to supply the want, which has long been a subject of complaint among those whom the citizens of Boston have charged with the general superintendence of their public schools, as well as with those who are appointed to the immediate instruction of them ; of a Book of Exercises in Reading and Speaking, better adapted than any English compilation that has yet appeared, to the state of society as it is in this country, and less obnoxious to complaint, on the ground of its national or political character, than it is reasonable to expect that any English compilation would be, among a people whose manners, opinions, literary institutions, and civil government, are so strictly republican as our own.

Extract from the Records of the School Committee, Boston.

At a meeting of the School Committee, held July 18, 1823, it was ordered, that the American First Class Book be hereafter used in the public reading schools instead of Scott's Lessons.

Attest, WILLIAM WELLS, *Secretary.*

The " American First Class Book" which has been favourably known to the public for several years, was intended, as its name imports, for the most advanced classes of the highest Schools, in which reading forms a part of the course of instruction. The extensive and increasing circulation, which that valuable selection has received and is receiving, and the success with which the use of it has been

1

attended, are sufficient indications that such a book was needed, and that the Author has made a judicious selection and arrangement of exercises.—*American Journal of Education.*

The AMERICAN SPEAKER, or Exercises in Rhetorick, being a new and copious selection of Speeches, Dialogues, and Poetry, from the best American and English sources, suitable for Recitation. Price $1,25.

The obvious want of interesting and *modern* extracts for the use of schools in which Rhetorick is taught, has produced the present compilation. Although several old and approved pieces are retained, it may be said with truth that this is a *new* selection, embodying the best of what has heretofore been published and much which has never before appeared in any school book. The friends of eloquence will be gratified to possess so many brilliant extracts in so small a compass, and the American patriot will be glad of an opportunity to compare the eloquence of his countrymen with that of the mother country.—*Preface.*

This Compilation, of which Mr. Fowle is only the editor, contains a better selection of modern pieces, and particularly *dialogues* than any similar book extant. The American Journal of Education, whose editor is a distinguished Rhetorician, says, " The American Speaker is a book which we are glad to see; it adds much that is new and interesting to the previous stock, and all the pieces possess that vivacity of character, which is a great point in producing animated delivery—the very soul of good speaking."

EASY LESSONS in GEOGRAPHY and HISTORY, by *Question and Answer,* designed for the use of the younger classes in the New England Schools. Second edition, revised and improved; to which are prefixed, The ELEMENTS OF LINEAR DRAWING. By JOSEPH ALLEN, Minister of Northborough, Mass. Price 12½ cents.

This is one of the happiest attempts for the improvement of primary education that has fallen under our notice. The arrangement of the materials is exceedingly judicious; being managed so as to lead the young mind gradually through a natural and easy and interesting succession of thought in which the elements of Geography and national history are very finely combined. No mechanical process of memory is employed; all is rendered intelligible and familiar, and at the same time equally instructive and pleasing. Primary education has been very deficient hitherto in aids such as this. We would earnestly entreat the attention of School Committees to this practical and useful work. Vastly more may be done with young children, than merely teaching them to spell and read: and books such as this, in the hands of attentive teachers, might be rendered as much a matter of recreation as of study; whilst a large portion of time now mispent would be redeemed for the invaluable purposes of early improvement.—*Journal of Education, Vol. II. No. 7.*

GREEK GRAMMAR, for the use of Schools, from the German of PHILIP BUTTMANN, edited by EDWARD EVERETT. Second edition. Price $2,00.

The deficiency of the Greek Grammars in use in this country, has been generally felt and loudly complained of. Under these circumstances the translator (Prof. E. EVERETT) was led to prepare a translation of the most approved of the Greek Grammars in use in Germany. It is well known that the Germans have paid a greater attention to philological pursuits than any other people. As a philosophical and practical grammarian, Prof. BUTTMANN, of the University of Berlin, is allowed by his countrymen to hold the first rank. He published three Greek Grammars, of which the smallest is here presented to the American scholar in a translation. It passed through many editions in Germany in a short time; and the rapid sale of the first edition of the translation, proves that the merits of the work and the value of the author's labors, are well appreciated in this country.

GREEK GRAMMAR, principally abridged from that of BUTTMANN, for the use of Schools. Price 62½ cents.

Preface.

The superiority of Buttmann's Greek Grammar over any other is acknowledged; but it appears to many instructors, whose judgment deserves the highest respect, that the work presupposes in those who are to make use of it more maturity of mind, than is to be expected of beginners. A desire has, therefore, been repeatedly expressed, that a small Grammar, in accordance with Buttmann, might be prepared for those entering on the study of the Greek language. Such a grammar is now offered to the practical teacher.

This abridgment is designed to contain only the accidence and first principles of the language. All matter that is not of immediate importance and utility has been rejected; and it has uniformly been endeavoured to unite simplicity in the arrangement with clearness and conciseness in the expressions. In preparing the work, the best school grammars of the Germans and the English have been carefully consulted on every point, and the judgment of the editor in what is retained and what is omitted has been directed by a comparison of the best materials. Particular assistance in these respects has been derived from the smaller grammar of Thiersch.

The practical instructer has here in a small compass all that is essential to be taught in preparing a pupil for any of our colleges.

The FRENCH PHRASE BOOK, or Key to French Conversation. Containing the chief Idioms of the French Language. By M. L'ABBE BOSSUT. Price 37½ cents.

The Editor feels no hesitation in asserting that after students have perfected themselves in the contents, even of this small Tract, they will have no difficulty in reading any French book, as far as depends on the peculiar idiom or construction of the language. By learning these familiar and idiomatic phrases, the young English scholar will acquire the French language and idiom exactly in the same manner as it is acquired by a native—by practice and example and not by rule. Rules are not to be despised; but they are rather adapted to perfect than to initiate.

Cummings' Elementary Works.

An INTRODUCTION to ANCIENT and MODERN GEOGRAPHY, to which are added Rules for Projecting Maps, and the Use of Globes. Accompanied with an Ancient and Modern Atlas. By J. A. CUMMINGS. Tenth edition, revised and improved. Price of Geography, 62½ cents. Price of Modern Atlas, 75 cents. Price of Ancient Atlas, 87½ cents.

The very liberal patronage which has been given to this work in its original form, has imposed on the proprietors an obligation to improve it as much as possible. It is confidently believed that the public will be satisfied that this obligation has been faithfully fulfilled in the present edition. The work is considerably reduced in size, by excluding such tables and abstract statements as are uninteresting and unimportant in an elementary treatise; but it contains more than the preceding editions of such matter as is useful to children.

In Cummings' Geography Improved, the questions are placed at the end of the several chapters. This is more convenient for the scholar than the former arrangement. Instead of adding a pronouncing vocabulary at the end of the book, most of the difficult names have their true pronunciation given where they occur; this will be found a very valuable improvement, and it is peculiar to this Geography.

A great number of cuts, very neatly engraved, ornament the work, and tend to illustrate the subjects, and render them interesting.

The simplicity of style and interesting manner of description, by which this work is characterized, have enabled it to sustain a high rank, and secured it a very extensive circulation. It is not to be forgotten that the public are indebted to Mr. Cummings for the general system of instruction in this science which now prevails, and which has been found so useful. The editor of the present edition has endeavored to retain the peculiar excellences of the original work; to correct its errors; and to make such improvements as will render it worthy of a still more extensive patronage. A great part of the work has been newly written, and it is interspersed with such instructions to scholars and teachers, as will facilitate the study of it, and render it permanently useful.

The ATLAS for the Improved Edition is newly engraved, contains a chart showing the comparative height of the principal Mountains, and of the comparative length of the principal Rivers in the world, and is intended to be as perfect as a work of the kind can be made.

CUMMINGS' PRONOUNCING SPELLING BOOK. Price 25 cents.

The extensive sale of this work, and the numerous testimonies of instructers and literary gentlemen, are sufficient proof of the excellence of its plan and execution. Indeed, those who consider the importance of teaching their children the correct pronunciation of the English language, while they are learning to read it, cannot but highly appreciate the plan of this Spelling Book. How frequently do we find that the errors in pronunciation, into which persons are allowed to fall in their childhood, continue to be repeated through life. It is certainly much easier for a child to acquire a correct pronunciation, than for an adult to reform a bad one.

In using Cummings' Spelling Book it requires but little pains to render the child able to determine the precise sound of every letter, and to make it more natural and easy for him to pronounce the words correctly than incorrectly. A little embarrassment is experienced at first, from the small letters which are used to designate the sounds of the others, but this is readily overcome, and the scholar is then possessed of a system which will enable him to pronounce all the words in his book correctly, and the instructer is saved the labor and frequent interruption which are suffered by the necessity of pronouncing words for the scholar.

In this edition a selection of very interesting reading lessons has been added, making it, it is believed, altogether the best Spelling Book in use.

CUMMINGS' FIRST LESSONS IN GEOGRAPHY and Astronomy. Price 25 cents.

It is hardly necessary to say any thing in commendation of a work which is so extensively known, and so generally esteemed.

The public are not, however, sufficiently aware of the ease and advantage with which such simple lessons, in these important sciences, may be learned by small children. The time which is nearly wasted in the study of Grammar, if employed in acquiring the elements of more exact sciences, would give the scholar not only a taste for them, but important information. Geography is so simple a science, that children of six or seven years of age may begin to understand it; and when a few of its elements are acquired, something may also be profitably taught them of the worlds around us.

Cummings' First Lessons is known to be far preferable to any other work in use, for introducing these subjects to the minds of children. The proprietors have taken great pains to render the work correct, and deserving of a still more extensive patronage.

The NEW TESTAMENT of our Lord and Saviour Jesus Christ, with an Introduction giving an account of Jewish and other sects; with Notes illustrating obscure passages, and explaining obsolete words and phrases; for the use of Schools, Academies, and Private Families. By J. A. Cummings, Author of Ancient and Modern Geography. Second edition. Revised and greatly improved. Stereotype edition. Price 75 cents.

·CUMMINGS' QUESTIONS on the New Testament, for Sabbath Exercises in Schools and Academies, with four Maps of the countries through which our Saviour and his Apostles travelled. Price 37½ cents.

Colburn's Works.

COLBURN'S FIRST LESSONS, or Intellectual Arithmetic, upon the Inductive Method of Instruction. By Warren Colburn. A. M. Stereotype Edition. Price of Book 37½ cents. Price of Plates 12½ cents.

The merits of this little work are so well known, and so highly appreciated in Boston and its vicinity, that any recommendation of it is unnecessary, except to those parents and teachers in the country, to whom it has not been introduced. To such it may be interesting and important to be informed, that the system of which this work gives the elementary principles, is founded on this simple maxim; that, *children should be instructed in every science, just so fast as they can understand it.* In conformity with this principle, the book commences with examples so simple, that they can be perfectly comprehended and performed mentally by children of four or five years of age; having performed these, the scholar will be enabled to answer the more difficult questions which follow. He will find, at every stage of his progress, that what he has already done has perfectly prepared him for what is at present required. This will encourage him to proceed, and will afford him a satisfaction in his study, which can never be enjoyed while performing the merely mechanical operation of cyphering according to artificial rules.

This method entirely supersedes the necessity of any rules, and the book contains none. The scholar learns to reason correctly respecting all combinations of numbers; and if he reasons correctly, he must obtain the desired result. The scholar who can be made to understand how a sum *should* be done, needs neither book nor instructer to dictate how it *must* be done.

This admirable elementary Arithmetic introduces the scholar at once to that simple, practical system, which accords with the natural operations of the human mind. All that is learned in this way is precisely what will be found essential in transacting the ordinary business of life, and it prepares the way, in the best possible manner, for the more abstruse investigations which belong to maturer age. Children of five or six years of age will be able to make considerable progress in the science of numbers, by pursuing this simple method of studying it; and it will uniformly be found that this *is* one of the most useful and interesting sciences upon which their minds can be occupied. By using this work children may be farther advanced at the age of nine or ten, than they can be at the age of fourteen or fifteen by the common method. Those who have used it, and are regarded as competent judges, have uniformly decided that more can be learned from it in one year, than can be acquired in two years from any other treatise ever published in America. Those who regard economy in time and money, cannot fail of holding a work in high estimation which will afford these important advantages.

Colburn's First Lessons are accompanied with such instructions as to the proper mode of using them, as will relieve parents and teachers from any embarrassment. The sale of the work has been so extensive that the publishers have been enabled so to reduce its price, that it is, at once, the cheapest and the best Arithmetic in the country.

COLBURN'S SEQUEL to INTELLECTUAL ARITHMETIC, upon the Inductive Method of Instruction. Price $1,00.

This work consists of two parts, in the first of which the author has given a great variety of questions, arranged according to the method pursued in the First Lessons; the second part consists of a few questions, with the solution of them, and such copious illustrations of the principles involved in the examples in the first part of the work, that the whole is rendered perfectly intelligible. The two parts are designed to be studied together. The answers to the questions in the first part are given in a Key, which is published separately for the use of instructers. If the scholar find any sum difficult, he must turn to the principles and illustrations, given in the second part, and these will furnish all the assistance that is needed.

The design of this arrangement is to make the scholar understand his subject thoroughly, instead of performing his sums by rule.

The First Lessons contain only examples of numbers so small, that they can be solved without the use of a slate. The Sequel commences with small and simple combinations, and proceeds gradually to the more extensive and varied, and the scholar will rarely have occasion for a principle in arithmetic which is not fully illustrated in this work.

KEY to COLBURN'S SEQUEL. Price 75 cents.

COLBURN'S INTRODUCTION to ALGEBRA, upon the Inductive Method of Instruction. Price $1,25.

Those who are competent to decide on the merits of this work, consider it equal at least, to either of the others composed by the same author.

The publishers cannot desire that it should have a higher commendation. The science of Algebra is so much simplified, that children may proceed with ease and advantage to the study of it, as soon as they have finished the preceding treatises on arithmetic. The same method is pursued in this as in the author's other works; every thing is made plain as he proceeds with his subject.

The uses which are performed by this science, give it a high claim to more general attention. Few of the more abstract mathematical investigations can be conducted without it; and a great proportion of those, for which arithmetic is used, would be performed with much greater facility and accuracy by an algebraic process.

The study of Algebra is singularly adapted to discipline the mind, and give it direct and simple modes of reasoning, and it is universally regarded as one of the most pleasing studies in which the mind can be engaged.

KEY to COLBURN'S ALGEBRA. Price 75 cents.

CORNELIUS NEPOS, de vita Excellentium Imperatorum. From the third edition of J. H. BREMI. With English Notes. Price 75 cents.

Nepos is, more than any other Roman writer, suited to be put into the hands of boys, who have made sufficient progress to be able to read a Roman author in

1*

course. The simplicity and classical character of his style, the separate lives, full of interest and not long enough to weary, the extent of history, of which he gives a pleasing outline, by presenting as in a gallery those illustrious men who directed the fortunes of antiquity, the general purity of the moral tendency of his writings, and the favorable moral influence which always follows from the true history of great men, are circumstances which sufficiently explain why he is so universally adopted in the European Schools, and is beginning to be introduced in so many of our own.

The few notes which accompany this edition are selected and abridged from the commentary of BREMI. In some instances the phraseology of Bradley, an English editor, has been adopted, where his remarks coincided with those of the continental editor. The notes would have been selected much more freely but for the fear of making the volume too large. They almost all of them relate to questions of grammar and language. These are the points, to which the attention of boys is to be directed.

In Press. An ELEMENTARY TREATISE on MINERALOGY, and Geology, designed for the use of pupils,—for persons attending Lectures on these subjects,—and as a Companion for Travellers in the United States of America. Illustrated with Plates. By PARKER CLEAVELAND, Professor of Mathematics and Natural Philosophy, and Lecturer on Chemistry and Mineralogy, in Bowdoin College. Third Edition in two volumes.

This work is now extensively known and used in the United States, and has been received with high approbation in Europe. The general plan of this edition is the same as that of the second; but the work is enlarged by the introduction of new species of minerals and new localities. Great efforts have been made to obtain correct descriptions of the localities of American minerals; and more especially to furnish accurate information concerning those minerals, which are employed in the useful and ornamental arts. Although the mineral riches of the United States have been but imperfectly investigated, yet sufficient is already known to show their importance in regard both to the wealth of individuals and the public good.

A CATECHISM of the ELEMENTS of RELIGION and MORALITY. By Rev. WILLIAM E. CHANNING.

The first object which the writer of this Catechism has had in view, has been to present to the minds of Children the great elementary principles of moral and religious truth, with the utmost possible simplicity of language.

The CHILD'S COMPANION; being an easy and concise Reading and Spelling Book, for the use of Young Children. By CALEB BINGHAM, A. M. Price 12½ cents.

Few men have attained so high eminence as a successful Instructer and Compiler of School Books as Mr. BINGHAM. Though published many years ago, his books still retain their place in many of our schools; and where they have been displaced by more recent compilations, their place has been often supplied by works of far inferior merit—this remark is especially true as applied to the Child's Companion. For simplicity and adaptation to the comprehension of quite young children, and at the same time for truly philosophical arrangement, this work yields to none of the kind in the English language. The steps from the most simple to the more complicated words and sentences, is so easy and natural that the child is brought to master the most difficult without great effort, and above all without disgust.

M. T. CICERONIS ORATIONES Selectæ, Notis Anglicis Illustratæ. Editio Quarta. 12mo. Price $1,50.

The merits of this book, as originally prepared for the use of Phillips Exeter Academy, are sufficiently known to the public. In this new edition, it has been the principal object of the editor to exhibit a better text than has hitherto been given in the school editions of Cicero, and by a more careful punctuation to place the meaning of the Author in a clearer light. The English Notes have most of them been retained, and placed at the end of the volume. They have however received many corrections and additions; and particularly Voel's Analytical and Synoptical Tables have been prefixed to the Notes of each Oration, showing the object of the orator and the course of the argument. These supersede the necessity of sets of Questions, as they suggest them to the instructer and pupil. On the whole it is believed that the value of this book has been essentially increased in this edition, and that little remains to be desired in this portion of the Latin course pursued in our schools.

Will speedily be Published,

Cubi's System of Translation.

Applied to the French,—LE TRADUCTEUR FRANÇOIS, or a new and practical system for translating the French language. By MARIANO CUBI I SOLER. Second edition, corrected, revised, and much improved.

Applied to the Latin,—The LATIN TRANSLATOR, or a new and practical system for translating the Latin language.

Applied to the Greek,—The GREEK TRANSLATOR, or a new and practical system for translating the Greek language.

Applied to the English, for Spaniards,—The ENGLISH TRANSLATOR, ó nuevo i práctico sistema de traduccion, para los que hablan español. Por MARIANO CUBI I SOLER.

Until now, Grammars and common Dictionaries have been the only auxiliaries which students have enjoyed in prosecuting the study of a foreign language. The intricate idioms and delicate subtleties of expression, have been left to the oral explanation of the tutor ; and the proper names as well as the grammatical niceties to the supposed historical or philological knowledge of the student. It is evident, that, as it is within the power of very few learners to command the continued attention of an instructer, and of still fewer, to obtain a profound knowledge of Grammar without an acquaintance with the language it treats, the progress of the majority must have been and is much retarded, or attended with many great, often insurmountable, difficulties.

To avoid all these inconveniences, by offering speedy success to the student, as the certain reward for his exertions, is the chief design of the author, in the system of translation, which he now offers to the public. Being circumscribed by the limits of a mere introductory notice, he will briefly state, that to accomplish his end, a collection of classic pieces, written in the language to be taught, is made, and arranged according to gradual difficulty. Notes explanatory of every intricacy of idiom, or nicety of Grammar are given at the bottom of every page, as these obstacles occur. At the end of this selection, thus arranged, and thus commented, a vocabulary is found in which *every word without exception*, whether proper or appellative, primitive or derivative, simple or modified, is fully analyzed, and its signification accurately explained.

Hence it is apparent, that if the meaning of every word simply, or of two or more combined into an idiom, be placed within the comprehension of the student, the sense of a whole paragraph or page, cannot remain for a long time obscure. These advantages will not, as many might, at one glance, suppose, offer unnecessary facilities to the learner. He cannot, unless it be through the medium of study and reflection conceive the meaning of any sentence, as it has, in no instance, been *conceived for him.* He may find facilities to attain this object soon and without despair, but it requires study and constant attention. Industry finds, in short, every incentive for its exercise, as no obstacles are presented which render it useless. The author may now speak with that confidence with which practical experience inspires, as this system has already been successfully applied to the French and Spanish languages.

How far this new mode of translation has advantages over the common way now pursued in teaching foreign idioms; and how far it has claims, if any, to originality, will be shown and clearly demonstrated in a pamphlet which the author is now preparing, and which will very soon be published, wholly devoted to this subject.

ELEMENTARY CATECHISM on the CONSTITUTION of the UNITED STATES, for the use of Schools. By ARTHUR J. STANSBURY. Price 37½ cents.

How small a portion of the citizens of this Republic have even a tolerable acquaintance with their own Constitution? It appears that this culpable want of acquaintance with what is of such deep interest to us all, is to be traced to the omission of an important part of what ought to be an American education, viz. the study of the civil institutions of our country. The foregoing work has been prepared with a view to such an experiment. It is written expressly for the use of boys, and it has been the aim and effort of the writer to bring down the subject completely to a level with their capacity to understand it. Whether he has succeeded the trial must show. He has purposely avoided all abstruse questions, and has confined himself to a simple commonsense explanation of each article.

DELECTUS SENTENTIARUM GRÆCARUM, ad usum tironum accommodatus : cum Notulis et Lexico. Editio Americana tertia prioribus emendatior. Price 62½ cents.

If the popularity of a book be an evidence of its having attained its object, the Greek Delectus has been eminently successful. Its merit consists in its simplicity, clearness and precision, by which, with a familiarity with his Grammar, the scholar may make great progress, relieved at once of useless labor, and yet compelled to habits of faithful study and thorough discipline. The Publishers have judged, that, where approbation has been so decidedly expressed, it would be an unwise attempt to substitute a better book; and that they could perform no more acceptable service, than to continue the present work, in as perfect a form as possible. The third American edition has, accordingly, been revised with care. The Notes have been considerably enlarged, critical peculiarities both in Etymology and Syntax pointed out, and a comparison instituted, in many cases, between the Greek and the Latin. The Lexicon is made to embrace not only all the words occurring in the Text, but likewise the irregularities of Tense in each Verb are prominently stated, and the quantity of the doubtful vowels is also marked in conformity to Morrell's Thesaurus.

This work is now used in the Boston Latin School.

ENFIELD'S INSTITUTES of NATURAL PHILOSOPHY, Theoretical and Practical, with some corrections ; change in the order of the branches ; and the addition of an Appendix to the Astronomical part, selected from Mr. ERVING's Practical Astronomy. By SAMUEL WEBBER, A. M., A. A. S. Fourth edition, with improvements. With Plates, in 1 vol. Quarto. Price $7,50.

FLORULA BOSTONIENSIS. A Collection of Plants of Boston and its vicinity, with their generic and specific characters, principal synonyms, descriptions, places of growth, and time of flowering, and occasional remarks. By JACOB BIGELOW, M. D. Professor in Harvard University. Member of the Linnæan Societies of London and Paris. Second edition greatly enlarged. To which is added a Glossary of the Botanical Terms employed in the work. 1 vol. 8vo. Price $2,75.

The first edition of the Florula Bostoniensis was published in 1814, for the use of a Botanical Class in this city. It was intended to contain intelligible descriptions of the more common and interesting plants found within a circuit of about ten miles around Boston. Its publication was at that time rendered necessary by the great deficiency of books relating to American plants, and by the difficulty of obtaining foreign works of a character suited to supply this deficiency. The edition now offered to the public contains about twice the number of plants which were included in the first edition. Many of the former descriptions have been enlarged or amended from re-examination of living plants, and many have been written out anew. Although the work more immediately applies to Boston and its environs, yet I have inserted in this edition all such plants as I have formerly collected and described in any part of the New England States. For the convenience of students a Glossary, explanatory of the technical terms used in the work, is added to this edition.

An ELEMENTARY COURSE of CIVIL ENGINEERING, translated from the French of M. I. SGANZIN, Inspector General of Bridges, Roads and Naval Depots, late Professor in the Royal Polytechnic School, Officer in the Legion of Honor, and Knight of the Royal Order of Saint Michael. From the third French edition, with Notes and Applications adapted to the United States. 1 vol. 8vo. With plates. Price $2,00.

The object of the translator in presenting this work to the public is to do something to supply what seems to him a great deficiency in the books on practical science in this country. He is acquainted with no work in English, which contains within a small compass, and in a form intelligible to common readers, those elementary principles of Engineering, which relate to building in stone, brick, or wood, and making roads, bridges, canals, and rail ways. Nearly all the books to be found on these subjects are suited only to the professed Engineer, and are either too voluminous, or too much involved in mathematical language to be accessible or intelligible to the greater part of learners and *practical mechanics.* The work of SGANZIN, of which he now offers a translation, seemed better suited than any other to the object he had in view. It has long had a high reputation in France, and has been used as a text book in the department of Civil Engineering at the Royal Polytechnic School in Paris ever since it was written.

In its present form the translator hopes it will be found useful not only to the professed student of Civil Engineering, but to the *practical mechanic,* and all persons engaged in any kind of building, in forming a road or rail way or digging a canal.

This translation is adopted at the United States Military Academy at West Point.

FROST'S ENGLISH PARSING EXERCISES. Five hundred Progressive Exercises in Parsing. Adapted to Murray's and other approved Treatises of English Grammar. By JOHN FROST. Price 12½ cents.

These Exercises, are carefully digested and arranged, so that the pupil learns how to manage one part of speech and one principle of Syntax, before he proceeds to others. The sentences illustrating each rule are distinctly classed, the difficulties which arise from the omission of a given point of speech or from a particular species of inversion, are separately pointed out and illustrated, and each important principle of Grammar thus becomes forcibly impressed on the youthful mind in association with several familiar examples.

From the American Journal of Education.

These Exercises will be found of great assistance in training children to accuracy and fluency in parsing. The language selected is mostly familiar; and the words of every lesson, therefore, are better adapted to the capacity and progress of young pupils, than is the case in exercise books which contain abstract sentiments and formal phraseology.

The FOUR GOSPELS of the NEW TESTAMENT, *in Greek,* from the text of GRIESBACH, with a Lexicon in English, of all the words contained in them : designed for the use of schools. Price $2,25.

Advertisement. This edition of the Four Gospels has been prepared in consequence of the new arrangement of the studies in Greek, preparatory to admission in the University at Cambridge. The Corporation have substituted the Boston edition of JACOB'S Greek Reader and the Four Gospels for the Collectanea Graeca Minora, and the whole of the New Testament. It has been deemed expedient to publish a separate edition of the Gospels. The text used is that of GRIESBACH, with the omission of the marginal readings, as not being appropriate to a School Book. A Lexicon of all the words in the Four Gospels, prepared with great care by a gentleman highly qualified for the task, is subjoined. It is hoped that the execution of the work will be found such as to merit the approbation of instructers and render it useful to learners.

The FRIEND of YOUTH, comprising a great variety of useful and interesting lessons in Prose and Poetry, adapted to the use of schools. By NOAH WORCESTER, D. D. Second edition. Price 75 cents.

The peculiar excellencies of this work consist in the purity and simplicity of the style and sentiments. In the Friend of Youth the beauty and simplicity of nature have been carefully regarded, while a pleasing variety has been preserved.

But the principal object of the author seems to have been to render the work totally destitute of such expressions and sentiments as flow from the corrupt passions of men, and engender discord and strife. It is not too much to say, that in this respect, this book is eminently distinguished from most of those now in use. If any Christian will keep in mind, that love to our fellow men is our first duty as social beings, and compare the amiable spirit, and the just and benevolent precepts which abound throughout this work, with the selfish and contentious effusions of selfish and jarring statesmen, of warring heroes, and of licentious poets, which so frequently disgrace the pages of others, we think he cannot hesitate in deciding which will afford him most aid in training up his children in the way they should go.

Prof. Farrar's Mathematics.

An ELEMENTARY TREATISE on ARITHMETIC, taken principally from the Arithmetic of S. F. LACROIX, and translated from the

French with such alterations and additions as were found necessary in order to adapt it to the use of American students. By JOHN FARRAR, Professor of Mathematics and Natural Philosophy in the University at Cambridge. Third edition, corrected and somewhat enlarged. 8vo. Price $1,00.

This is a philosophical treatise, in which the rules are all strictly demonstrated, a recommendation which no other Arithmetic published in this country is known to possess.

"It is important to remark that the Arithmetic will be of little advantage to any who are determined not to take the trouble of thinking and who have nothing of the spirit of inquiry and investigation. At the same time the book is calculated to awaken this spirit."—*Review of Cambridge Mathematics, Silliman's Journal.*

An INTRODUCTION to the ELEMENTS of ALGEBRA, designed for the use of those who are acquainted only with the first principles of Arithmetic. Selected from the Algebra of EULER. Second edition. By JOHN FARRAR, Professor of Mathematics and Natural Philosophy. 8vo. Price $1,50.

Amid the multitude of more recent treatises there are few in which the learner will find more interest and satisfaction than in this.

"Of EULER it is not necessary to say much to those who are in any degree acquainted with mathematical science. In clearness and elegance of demonstration and illustration he stands the prince of mathematicians, and in fertility of invention he has never been surpassed."—*Review of Cambridge Mathematics, Silliman's Journal.*

ELEMENTS of ALGEBRA. By S. F. LACROIX. Translated from the French for the use of the students of the University of Cambridge in New England. By JOHN FARRAR, Professor of Mathematics and Natural Philosophy. Second Edition. 8vo. Price $1,50.

This work comprehends many things not to be found in EULER. It has been generally preferred in the *French* schools to all other treatises.

"LACROIX appears to have been governed in preparing his mathematical works by the following principles; to give a demonstration as rigorous as the nature of the case would admit of every rule and principle of which any use is made. This is very different from the course pursued in most *American* and *English* books upon mathematics. In our treatises upon Arithmetic and Algebra, with a very few honorable exceptions, the rules are given in a very concise and purely didactic form, and whatever attempt there is at an investigation of them is thrown into notes which are seldom much consulted. Nor is the student generally to blame for not consulting them, as they are usually so ill adapted to the state of his knowledge that he finds it impossible to understand them."—*Review of Cambridge Mathematics, Silliman's Journal.*

ELEMENTS of GEOMETRY. By A. M. LEGENDRE. Member of the Institute and the Legion of Honor, of the Royal Society of London, &c. Translated from the French for the use of the students of the University at Cambridge, New England. By JOHN FARRAR, Professor of Mathematics and Natural Philosophy. Second edition, corrected and enlarged. 8vo. Price $2,00.

This is universally allowed to be the best and most complete treatise on the elements of *Geometry* that has yet appeared.

"On the Geometry of solids or volumes, the elements of LEGENDRE and LACROIX are very much more complete than those of EUCLID. On this point it is impossible to convey an adequate idea to those who are not to a considerable extent acquainted with the subject. - - Those who are only acquainted with the Geometry of solids or volumes as given by the older writers, we are sure will be surprised and delighted at the luminous and novel manner in which this part of elementary Geometry is exhibited. After what has been said it is scarcely necessary to observe, that *American* mathematical science is under great obligations to the translator for giving LEGENDRE's elements in so handsome an English dress. The translation is faithfully executed and accurately printed."—*Review of Cambridge Mathematics, Silliman's Journal.*

An ELEMENTARY TREATISE on PLANE and SPHERICAL TRIGONOMETRY, and on the application of Algebra to Geometry:

from the Mathematics of LACROIX and BÉZOUT. Translated from the French for the use of the students of the University at Cambridge, New England. Second edition. 8vo. Price $1,50.

The treatises upon Plane and Spherical Trigonometry, though concise, are abundantly sufficient for all the ordinary purposes to which they are applied. The portion on Conic Sections developes the leading properties of these curves in a very plain and satisfactory manner.

Although the analytical method is adopted it will be found to be attended with little or no difficulty. The immense advantages which it gives over the geometrical cannot but be perceived by the diligent and faithful student. New steps are supplied and all the aid to be derived from frequent references, are afforded, that the less experienced learner may proceed with as much dispatch as the nature of the subject will admit. The formulas in Trigonometry and Conic Sections which will be sought in vain in the common treatises will be found to be of the greatest importance in the higher parts of Natural Philosophy and especially in Astronomy.

. An ELEMENTARY TREATISE on the Application of Trigonometry to Orthographic and Stereographic Projection, Dialling, Mensuration of Heights and Distances, Navigation, Nautical Astronomy, Surveying and Levelling; together with Logarithmic and other Tables; designed for the use of the students of the University at Cambridge, New England. 8vo. Price $2,00. (This work is lettered Topography.)

The several parts of this volume are prepared with great care; the best English and French works were consulted and no pains spared in condensing into a small compass what was deemed most important relating to the topics here treated. The Logarithmic tables are from the stereotype plates of BOWDITCH's Practical Navigator, the correctness of which is too well known to need any recommendation.

FIRST PRINCIPLES of the DIFFERENTIAL and INTEGRAL CALCULUS, or the Doctrine of Fluxions, intended as an Introduction to the Physico-Mathematical Sciences; taken chiefly from the Mathematics of BÉZOUT. And translated from the French for the use of the students of the University at Cambridge, New England. 8vo. Price $1,50.

This will be found far more easy and satisfactory than the English treatises upon this subject. It was selected on account of the plain and perspicuous manner for which the author is so well known, as also on account of its brevity and adaptation in other respects to the wants of those who have but little time to devote to such studies.

An ELEMENTARY TREATISE on MECHANICS, comprehending the Doctrine of Equilibrium and Motion, as adapted to Solids and Fluids, chiefly compiled, and designed for the use of the students of the University at Cambridge, New England. By JOHN FARRAR, Professor of Mathematics and Natural Philosophy. 8vo. Price $4,00.

Under the term Mechanics, are comprehended, in this work, not only those topics which are usually treated under this name, but also Hydrostatics and Pneumatics. The leading propositions are demonstrated with great strictness, and are derived one after another from a very few fundamental principles. There is throughout particular reference to the practical uses of the science.

ELEMENTS of ELECTRICITY, MAGNETISM, and ELECTRO-MAGNETISM, embracing the late Discoveries and Improvements, digested into the form of a Treatise; being the Second Part of a Course of Natural Philosophy, compiled for the use of the students of the University at Cambridge, New England. By JOHN FARRAR, Professor of Mathematics and Natural Philosophy. 8vo. Price $3,50.

Many phenomena and theories are made known in this treatise that the reader will seek in vain in the best English works on these subjects.

An EXPERIMENTAL TREATISE on OPTICS, comprehending the Leading Principles of the Science, and an explanation of the more important and curious Optical Instruments and Optical Phenomena, being the Third Part of a Course of Natural Philosophy, compiled for the use of the

12   School and Classical Books

Students of the University at Cambridge, New England. By JOHN FAR-
RAR, Professor of Mathematics and Natural Philosophy. 8vo.   Price $3,00.
This treatise is mostly confined to what is capable of being established and illustrated by experiment. In the Notes to this and the foregoing volume are collected
many recently discovered facts and principles that have not yet been embodied into
the form of a treatise.

An ELEMENTARY TREATISE on ASTRONOMY, adapted to
the Present Improved state of the Science, being the Fourth Part of a
Course of Natural Philosophy, compiled for the use of the students of the
University at Cambridge, New England. By JOHN FARRAR, Professor of
Mathematics and Natural Philosophy. 8vo.   Price $3,75.
This is a plain and familiar view of the subject. It is intended to be at once
popular and profound. A great part of it may be read without difficulty by persons little skilled in the pure Mathematics; other parts will require more attention
and study and somewhat more aid from the subsidiary sciences.

ELEMENTS of NATURAL PHILOSOPHY. By E. G. FIS-
CHER, Honorary Member of the Academy of Sciences of Berlin, Professor
of Mathematics and Natural Philosophy in one of the Colleges of the same
city, &c. &c. Translated into French, with Notes and Additions, by
M. BIOT, of the Institute of France; and now translated from the French
into English for the use of Colleges and Schools in the United States.
Edited by JOHN FARRAR, Professor of Mathematics and Natural Philosophy in the University at Cambridge, New England. 8vo.   Price $3,00.
This is an ingenious compend of Mechanical and Physical Philosophy that has
been much used and highly approved in the German and French schools.

The GREEK READER, by FREDERICK JACOBS, Professor of
the Gymnasium at Gotha, and Editor of the Anthologia. From the last
German edition, adapted to BUTTMANN'S Greek Grammar.   Second
Boston edition. 1 vol. 8vo. Price $2,25.
Extract from the North American Review.
The Greek Reader, having been compiled by one of the leading scholars of the
age, is prepared throughout in a pure and masterly manner; proceeds methodically
from the simplest combination of words to the common attic style; and is so composed, that while the rules of grammar are illustrated in easy succession, an outline is given of mythology, ancient geography, and Grecian history. It is used in
almost all the good schools in Germany, and has there gained a decided expression
of public opinion in its favor, as the best of the many similar works, which have
been produced by the scholars of that prolific country.
In regard to the American edition, the chief question concerns its accuracy; and
this quality it possesses in an eminent degree. As the Notes and Lexicon are in
English, it affords the means of learning Greek without the embarrassing interven-
tion of another foreign tongue. That it contains references to the American translation of BUTTMANN's Grammar, will make it the more valuable to those who possess that work, without diminishing its utility for those, who continue to use the
more ancient manuals.
Preface to the Second Boston Edition.
This edition of the Greek Reader has been prepared with great care. The Lexicon has been revised by a scholar of distinguished accuracy, and great pains have
been bestowed in inserting such words and significations of words, as had acci-
dentally been omitted in the first edition, and in making other improvements in
conformity with the suggestions of experienced instructers. In the typographical
execution of the work, the traces of a diligent and skillful revision of the press will
probably be perceived. The text has undergone a thorough revision, having been
collated with the German edition of this work, as well as compared with some approved edition of the several authors from whom the extracts are taken; and the
references to the Grammar have been accommodated to the second edition of
BUTTMANN's Grammar.
At a meeting of the Corporation of Harvard College,
Voted, "That this work be made use of in examining candidates for admission
into the University after the year 1826, instead of Græca Minora."

DEUTSCHES LESEBUCH für Anfänger, i. e. German Reader

for Beginners. Edited by Dr. CHARLES FOLLEN, of Harvard University. 1 vol. Price $1,25.

The design of this book is expressed in the Preface. 'It is intended to meet an urgent want of all those who are engaged either in teaching or studying the German language in this country. It introduces the learner to the master works of modern German literature, and furnishes the teacher with a large number of classical examples, to illustrate the rules and peculiarities of the language.' The book is divided into a prosaic, and a poetical part. The prosaic part contains, in chronological order, select pieces from the works of Lesing, Wieland, Herder, Engel, Göthe, Johannes, Müller, Schiller, Heeren, A. W. Schlegel, F. Schlegel, Wackenroder, Hardenberg, Tieck, and Hoffmann. The poetical part exhibits specimens of various kinds of poetry, by Schiller, Göthe, Bürger, Herder, Tieck, and Korner. · A sketch of the history of German literature from the earliest times to the present, is contained in the Preface.

"This is one of the pleasantest and best selections we are acquainted with, for the purpose of introducing a beginner to the knowledge of a foreign language. The object of it, as stated in the preface, is to give a collection of examples illustrative of the rules and peculiarities of the language, from works of acknowledged classical rank, and at the same time to afford the learner a foretaste of the modern German literature. This object is, we think, well attained ; and though a task of no very formidable nature, yet it is one not unworthy of the attention of the learned scholar who has prepared the book, and to whom we are indebted for contributing his efforts to increase the means of cultivating one of the most useful and important languages of the present day."—*North American Review.*

In Press. A PRACTICAL GRAMMAR of the German Language, by Dr. CHARLES FOLLEN, Instructer of the German Language at Harvard University, Cambridge. 12mo.

This work is calculated to serve as a guide to the teacher ; as well as to those who wish to study the German language without the aid of an instructer. The author has consulted the most important grammatical works which have heretofore been published, on the continent of Europe, and in England ; with particular reference to the grammars of Heinsius, Noehden, and Rowbotham. He has endeavored to comprise in his grammar all that seemed to him really useful for acquiring a practical knowledge of the German language ; without entering into too minute details, or indulging in idle speculations. Every rule of the language is illustrated by examples, and exercises for translating from German into English, and from English into German. The pronunciation is facilitated by an analysis of all the sounds which belong to the German language, and by marking the accent of every word which occurs in this Grammar.

GOULD'S VIRGIL, with English Notes, and a Key for Scanning. Publius Virgilius Maro. Bucolica, Georgica, et Æneis. Accedunt Clavis Metrica, Notulæ Anglicæ, et Quæstiones, nec non Index vocabulorum Uberrima. Cura B. A. GOULD. In Usum Scholæ Bostoniensis. 8vo. Price $3,50.

This edition of Virgil is printed without the usual *order of construction*, or *interpretation.* The use of these pernicious helps not only prevents the pupil from ever acquiring the power of reading with ease and pleasure without them, but it is utterly subversive of one of the principal objects of studying the language,—that mental discipline which is acquired by the practice of critical and exact analysis.

If the habit of reading independently of artificial assistance be once formed, the want of such assistance is not felt. And it is found by experience that boys who have never used an *order* or *interpretation*, read Horace and Juvenal, as readily as they do Cicero and Tacitus ; and even with more confidence ; since they are aided by Prosody in overcoming many doubts in poetry, which they have no means of solving in prose. The use of the *interpretation* is discontinued in the best schools both in England and in this country ; as is also the absurd custom of explaining by Latin notes, which boys do not understand.

To aid the scholar in overcoming the real difficulties in understanding this author, copious English notes are added at the end. A list of the verses most difficult to scan is subjoined, with the method of scanning each. A few questions are also added, which may expedite the labor of the teacher in ascertaining whether the pupil has been thorough in his preparations.

The work is published on a fine paper, and beautiful type; and is, altogether, far superior to any other edition of Virgil in use.

2

Extracts from the North American Review. No. 52.
It is printed with great neatness, in a type of sufficient size, producing well defined, well filled, well rounded letters, such as the eye may dwell upon without pain or weariness. We are acquainted with no edition, which, as regards typography, the accuracy of the text, and the correctness of punctuation, we should believe, will be read with more satisfaction. We rejoice to find this edition of Virgil excluding the *order of construction*, or the *interpretation*, which has so long disfigured our school Virgil, and other Latin poets. This interpretation, and translations into English of similar demerit, have often been the miserable crutches by which boys have limped their weary way through the Æneid, wholly unsuspicious that they were in company with one of the greatest poets of ancient or modern times.

·The notes are various in their kind; and not among the least frequent or useful are those of a philological character. The editor will not be accused of superfluity or prolixness in this part of his work; and his reasons for brevity are such as every person of similar experience will accept without hesitation.

National Gazette and Literary Register. July 13, 1827.
Mr. GOULD has rendered much service to the ends of classical education in this country, by his editions of Virgil and Adam's Latin Grammar, and his *Excerpta* from Ovid, with Notes and Questions. The Virgil, in particular, deserves to be widely known, from the peculiar correctness of the text, which is substantially Heyne's, the variations from that, the best extant, being slight, and such only as a careful collation of all the acknowledged authorities appeared to require.

Both the Virgil and the Ovid are printed not merely with remarkable accuracy, but in a handsome form. Whatever is well done in this way promotes, or supports the cause of classical literature and the good old system of instruction, which are assailed in public opinion by empirical speculations and schemes of "tricking short cuts and little fallacious facilities." Of the many contemporary innovations with regard to the communication of knowledge and the general culture of the mind, they are but very few which deserve to be styled improvements.

GOULD'S GRAMMAR:—ADAM'S LATIN GRAMMAR, with some Improvements, and the following Additions: Rules for the Right Pronunciation of the Latin Language; a Metrical Key to the Odes of Horace; a List of Latin Authors arranged according to the different ages of Roman Literature; Tables, showing the value of the various Coins, Weights, and Measures, used among the Romans. By BENJAMIN A. GOULD, Master of the Public Latin School, Boston. Price, bound, $1,00.

"*It must be remembered that if the Grammar be the first book put into the learner's hands, it should be the last to leave them.*"—Preface to BUTTMANN's Greek Grammar.

This edition is adopted by the University at Cambridge, Massachusetts, and is recommended to the use of those who are preparing for that Seminary.

Extracts from the Journal of Education.
Mr. GOULD has in this edition of the Grammar rendered to classical instruction one of the most valuable services it has hitherto received in this country.

His endeavors to promote a uniform and correct pronunciation of Latin are an important addition to the value of the Grammar.

The acquisition of a correct pronunciation should be an object of attention, as early as possible in the course of instruction.

With regard to the general merits of the rules we need say nothing, after stating that they are sanctioned by the authority of our most reputable literary institutions.

On one circumstance, connected with the improved edition of ADAM's Latin Grammar, we congratulate every instructer. The care taken to accent penultimate syllables seems likely to succeed in banishing the hideous mispronunciations with which our School and College exercises were, and are sometimes disgraced.

We cannot take leave of this valuable school book without expressing our warmest approbation of it, and our gratitude for the facilities which its editor has afforded our youth for their progress in a language which is essential to a full understanding of their own; which is the avenue to professional life, and to the highest and most honored spheres of public usefulness.

GOULD'S OVID, with English Notes. Excerpta ex scriptis Publii Ovidii Nasonis. Accedunt Notulæ Anglicæ et Quæstiones. In usum

Scholæ Bostoniensis. This selection embraces portions of the Metamorphoses, Epistles, and Fasti. 8vo. Price $1,25.

Extracts from the Preface.

In preparing this little volume from the writings of Ovid, great care has been taken to admit *nothing in the slightest degree indelicate,* or improper for the study of youth. One object has been to furnish examples of the different kinds of measure used by this polished and fascinating writer. It is not a little surprising that in the whole course of studies preparatory for, and pursued at our Colleges, not a verse of pentameter measure occurs. This is the more surprising, since, in addition to the frequency of its use, this kind of versification may be considered one of the most easy and graceful which the ancient poets used.   •

As this book is designed for a kind of introduction to fabulous history, the notes give a more full account of the subjects connected with the matter immediately under consideration, than might otherwise seem expedient.

The questions are designed to direct the student's attention to the subjects of the notes, as well as to those of the text; for a knowledge of the characters here introduced will genera facilitate a proper understanding of all subsequent studies in Latin and Greek. lly

Extracts from the United States Review and Literary Gazette, for August, 1827.

As in his Virgil, so in these selections from Ovid, Mr. Gould has rejected the *order of construction* and the *interpretation* and for reasons, we think, perfectly satisfactory. The Notes, too, [in English] are of the same judicious character as those in his Virgil. And besides those which are intended to assist the pupil merely in the business of interpreting the author, many of them are devoted to the explanations of names which occur so frequently in the Metamorphoses, and which have so much to do with the fabulous history of ancient times.   ı

The Questions annexed are well adapted to direct the pupil's attention to what is most worthy of his notice, and thus to fix in his memory many important facts in ancient mythology, and history, and geography.

We cannot close our remarks without expressing our obligations to Mr. Gould for this additional contribution to the cause of good learning. While we look not only with complacency, but with great satisfaction, upon the various useful enterprises that engage the busy world about us, we are delighted occasionally to greet the scholar, who comes to take the rising generation by the hand, and make them familiarly acquainted with the favorite poets of ancient times.

GOULD'S HORACE, with English Notes. Quinti Horatii Flacci Opera : accedunt clavis metrica et notulæ Anglicæ, Juventuti accommodatæ. Cura B. A. Gould. 21mo.

This edition of Horace has been prepared with much care for the use of young gentlemen at School and at College; and of course is free from all indelicacies. It is of the *duodecimo* form, and of a size convenient for use, with the notes at the end. The notes are concise, and adapted to the degree of information which the student is supposed to possess at that stage of his classical education when this author is usually studied. ↘ It is presumed that most students have become acquainted with the leading characters in classical history and mythology before they take Horace. For this reason the notes are chiefly confined to the illustration of the text, ı. e. the peculiarities of this author, and to such explanations of the manners and customs of the time, and of the characters introduced, as seem necessary to a right understanding of the poet's allusions.

GRAGLIA'S NEW POCKET DICTIONARY of the Italian and English Languages. With a compendious Elementary Italian Grammar, from the last London edition. (*Now Stereotyping.*)

The general approbation, with which the numerous editions of this valuable little Dictionary, have been received ;—the rapidity, with which *fourteen editions* of it have been exhausted in England ;—and its universal adoption by the instructers of the Italian language in this country, are a sufficient recommendation of the work and a proof of its superiority to others compiled on the same plan. "Several attempts," says the preface to the fourteenth London edition, "have been made to surpass this Dictionary; but it still keeps up its advantages and the last edition was sold in half the time of the preceding one."

The vocabulary is copious and various, the definitions accurate, many difficult phrases and peculiar idioms explained, and the most common poetical terms, which often occasion so much embarrassment to beginners, are introduced and defined. Numerous re-publications have, however, been made at London, without a proper attention to the correction of typographical errors—and in consequence, the accents are often misplaced, the words both Italian and English, so mis-spelled, as to produce important misrepresentations of the sense, and the genders, parts of speech, &c. incorrectly stated. In the American edition, these errors have been carefully corrected, the whole text examined by the folio edition of ALBERTIS, many definitions and idioms added, and the vocabulary itself enlarged by about *two thousand* of the most important words, omitted in the London copies.—To make it more useful, the texts of the classics most commonly read in the schools and by beginners in this country, as the Scelta of Goldoni, the Notti Romane of Verri, the Gerusalemma Liberata, the Tragedies of Monti, and the Italian Reader, compiled for the use of Harvard University, have been consulted, and several hundred words not contained in London editions have been added from them.

The. ITALIAN ' and ENGLISH PHRASE BOOK, or Key to Italian Conversation; containing the chief Idioms of the Italian Language. Improved from M. L'ABBE BOSSUT. Price 37½ cents.

This work is on the same plan with the French Phrase Book by BOSSUT—and is a complete Key to the conversational idioms of the Italian language; and when these idioms are once mastered, the whole language is easily attainable. It cannot fail to be eminently useful to beginners.

COLLECTANEA GRÆCA MINORA ; with Explanatory Notes, collected or written by ANDREW DALZEL, A. M. F. R. S. E. Professor of Greek in the University of Edinburgh. Sixth Cambridge edition ; in which the Notes and Lexicon are translated from the Latin into English. 1 vol. 8vo. Price $2,25.

Preface to the Sixth Cambridge Edition.

It has long been a complaint, that the notes of Collectanea Græca Minora, being written in Latin, were not so useful as they might be to beginners, for whose use they were prepared. In this edition, therefore, the notes and lexicon have been translated into English; so that the work may be used without any previous knowledge of the Latin language. So numerous are the words and idioms in Latin authors, which may be illustrated by a knowledge of the Greek language, from which they were borrowed, that no reasonable man can doubt that the Greek should be studied first.

In this edition a few notes have been added, particularly upon the most difficult part—the extracts from Tyrtæus. The text also has been diligently compared with the latest and best editions of the works, from which the extracts were made, belonging to the library of Harvard University ; and a few new readings have been introduced, which throw light on obscure passages. It is hoped, therefore, that those who have heretofore used and approved the work, will be still better satisfied with it, now that it is more free from errors, and more easy and instructive to young students.

COLLECTANEA GRÆCA MAJORA. Ad usum Academicæ Juventutis accommodata ; cum Notis philologicis, quas partim collegit partim scripsit ANDREAS DALZEL, A. M. &c. Editio quarta Americana, ex Auctoribus correcta, prioribus emendatior, cum Notis aliquot interjectis. Cantabrigiæ, Mass. E prelo Universitatis. Sumptibus HILLIARD, GRAY et Soc. Bibliopolarum, Bostoniæ. 2 vols. 8vo. Price $7,00.

Extracts from a Review of this Edition.

The best criterion by which to estimate the value of works designed to facilitate the purposes of education, is actual experiment. The present selections from Greek literature have been many years before the public, and have constantly been coming more widely into use. Of the first volume there have appeared in England and Scotland at least eight, we believe nine, several editions, and five or six of the second ; and in the United States, we have now the fourth edition of the whole work from the press of the University at Cambridge. A book, to meet with such success, must be well adapted to its end.

Of all the editions which have thus far appeared in Great Britain or America, we do not hesitate to pronounce this to be the most correct. It exhibits the clearest marks of indefatigable diligence and conscientious accuracy on the part of its learned and unassuming editor. Instead of vague and indiscriminating praise, we will endeavor to explain its peculiar advantages. Our account will be a short one, though the labors which we commemorate extended through years.

The chief object of the American editor, Professor JOHN S. POPKIN of Cambridge, was to make the book a correct one. It had gone through so many editions, and each new one had repeated so many of the errors of the last, and made so many of its own, that both the text and the notes had become very much disfigured. Not only accents and letters were often wanting, but words, and sometimes whole lines were omitted ; especially in the notes. In the third American edition, these were in a good degree amended ; in the fourth the same purpose has been most assiduously pursued. To do this the original sources of the notes and text were consulted, and these, together with other good editions of the several writers, were diligently compared. Not a few fractures and dislocations were repaired by means of an early edition of the Collectanea. When the sense was found broken and obscure, it appeared on examination that words, lines, and sometimes several lines had been omitted : particularly where a word was repeated at no great distance, the intervening words were sometimes passed over in printing.

We hope we have said enough to justify our preference of Professor Popkin's edition of the Græca Majora over any other. To give a more distinct idea of what he has accomplished, we venture to affirm, after a close computation which may be relied upon, that of errata in the copy greater and less, he has corrected as many as ten thousand. If after all his care and pains, he has made any or left any, they can be easily marked and corrected, as the present edition has been made on stereotyped plates. It was an undertaking of long and toilsome diligence to correct the press and the copy of a work of this kind, collected from so many sources, and referring to so many authorities.

Not less than five hundred volumes were of necessity consulted.

An EPITOME of GRECIAN ANTIQUITIES. For the use of Schools. By Charles D. Cleaveland. Price $1,00.

To the Publishers.

I received a few days since your letter of inquiry concerning the "Epitome of Grecian Antiquities," and am happy in the opportunity of expressing an opinion of that little work.—While in progress the plan received my full approbation ; and the diligence of the compiler in procuring and consulting all the proper authorities and the unusual care bestowed in superintending the publication led me to expect a well executed work. In this I have not been disappointed. Potter's Archæologia Græca is voluminous and expensive, and the works of Robinson and Bos have not been re-published in this country. Some work of the kind is absolutely necessary for the classical student in every stage of his progress. In this state of things Mr. Cleaveland's well digested manual supplies a deficiency and obviates an objection, which has heretofore existed, to making Grecian Antiquities a separate and particular study in our Academies and Classical Schools.

WILLIAM CHAMBERLAIN,
Professor at Dartmouth College.

In Press. GROVES'S GREEK LEXICON. A Greek and English Dictionary. By the Rev. John Groves. With additions, by the American editor.

The object of the compiler of this work (as stated in his Preface) was, to offer to the public a *Dictionary*, which young Greek scholars could use with ease and advantage to themselves; but sufficiently full to be equally serviceable as they advanced ; a book, that would answer for School, for College, and more particularly for reading the New Testament and the Septuagint.

In the arrangement of the words, the alphabetical method has been adopted, as best suited to the capacity and diligence of the young learner. Immediately after each word is placed its form of declension or conjugation, together with any peculiarity attending it, such as the attic genitive of contracted nouns, the future or perfect of any dialect peculiar to certain verbs, &c. Next is placed the derivation or composition of the word. The English significations of the Greek word follow next. In this part two or three synonyms have been given for each signification. After the significations are subjoined any irregularities or varieties arising from dialect, &c. and some of the more difficult inflexions of each word are added. A considerable number of new words have been introduced. These consist of words occurring in the authors usually read, and in the Septuagint, which are not to be found in Schrevelius. They are also taken from the Greek tragedies now generally read at schools, and from elementary books latterly introduced. All the inflected parts of words which are in Schrevelius are to be found in this Dictionary, with many others. These consist of oblique cases of nouns, pronouns, and participles ; of persons, tenses, moods, &c. of verbs. There is no English attached to these parts and inflexions, but a reference is given to the theme, where all the significations will be found. This part of the work has been particularly attended to

2*

where it respects the New Testament; and thus the work will be useful both to
beginners and to those persons who may wish to revive their knowledge of Greek,
especially of the New Testament; they will also find in this Dictionary all the
words occurring in the Septuagint. Every thing has been retrenched that did not
coincide with the young scholar's capacity; and he has here given to him what he
will notice at present, instead of what he may look for at a more advanced period
of his studies; he has here laid before him what will be immediately useful, for
what might be serviceable hereafter.

This American edition, in conformity with the plan of the work, will be improved
by the addition of considerable supplementary matter particularly adapted to the
wants of younger students, for whose use the work is principally designed.

Will speedily be Published. HOMER'S ILIAD, with English
Notes. In 2 vols. on the plan of GOULD's edition of Virgil, Ovid, Horace, &c.

ELEMENTS of LOGIC, or a Summary of the General Principles
and Different Modes of Reasoning. By LEVI HEDGE, LL. D. Professor
of Natural religion, Moral Philosophy, and Civil Polity, in Harvard University. *Stereotype edition.* Price 87½ cents.

Professor HEDGE made his Logic after twenty years experience in teaching the
science. His object was to form a system adapted to the present improved state
of intellectual philosophy. The Logics before in use were deficient in parts of primary importance. The instruction they furnish on the article of reasoning is
almost exclusively confined to the principles of sylogism. They contain very little
relating to moral evidence and those modes of reasoning by which the practical
business of life is carried on. The success which has attended the sale of HEDGE's
Logic is ample proof that it is suited to the wants of the community. It has su-
perseded those before used in most, if not all, the Colleges in New England and in
several of those in the middle, southern, and western states. It is also extensively
used in Academies and Schools of the higher order, in various parts of the United
States. In the different editions this treatise has been carefully revised by the
author, and in the fourth edition two chapters were added, containing the principles of controversy—and a body of rules for the interpretation of written docu-
ments.

JOHNSON'S DICTIONARY of the English Language, as im-
proved by TODD, and abridged by CHALMERS, with WALKER's Pronoun-
cing Dictionary combined ;—to which is added, WALKER's KEY to the
Classical Pronunciation of Greek, Latin, and Scripture Proper Names.
Edited by JOSEPH E. WORCESTER, A. M., A. A. S., &c. 1 vol. 8vo.
Price $5,25.

This Dictionary comprises a complete reprint of Mr. Chalmers's Abridgment;
Mr. Walker's Principles of English Pronunciation, his Pronunciation of all the
words found in his Dictionary, together with his Critical Remarks on the pronunciation of particular words; and also Walker's Key entire.

It possesses the following advantages, not found in the three works above
mentioned, of which it is chiefly composed.

1. The words added by Mr. Todd, exceeding fourteen thousand in number, are
discriminated from the rest.

2. Words pronounced obsolete by Dr. Johnson or Mr. Todd, are so marked.

3. Words designated by Dr. Johnson or Mr. Todd as primitive words, are
distinguished from such as are derivative.

4. Additional matter or remarks, etymological, critical, and explanatory, from
Dr. Johnson and Mr. Todd, are occasionally inserted.

5. To the many thousand words not found in Mr. Walker's Dictionary, the
pronunciation has been given according to his Principles.

6. With regard to the pronunciation of many words, respecting which other
orthoepists differ from Walker, and he has neglected to exhibit this difference,
the mode adopted by others is here given.

7. An Appendix, containing besides other matter, all the further additional
words, (about one thousand,) inserted by Mr. Todd in his second edition, published
in London in 1827.

8. Dr. Johnson's Preface to his folio edition, and Mr. Todd's Introduction.

9. An alphabetical list of the Authors referred to as authorities for the use of
words, with the time specified when they flourished.

ILLUSTRATIONS of PALEY'S NATURAL THEOLOGY, with Descriptive Letter Press. By JAMES PAXTON, Member of the Royal College of Surgeons, London. 1 vol. 8vo. Price $2,75.

"Of muscular actions, even of those well understood, some of the most curious are incapable of popular explanation, without the aid of Plates and Figures."— *Paley's Theology, Chap.* IX.

PALEY's Natural Theology has long been held in high estimation as a work eminently useful to young persons both for the quantity and clearness of the information it imparts, and the valuable deductions and conclusions which the author establishes in the progress of his work. But every person not previously conversant with the subject must have felt the want of some figures, some delineations for the eye, in order to comprehend fully the subjects treated of. These illustrations are beautifully executed in thirty six lithographic plates, large octavo size, with descriptive letter press; and render very clear the different mechanical functions of the bones, muscles, arteries, veins, viscera, &c. of animals, and many similar and curious operations in the vegetable kingdom. It is an indispensable accompaniment to one of the best books in the English language.

In Press. A GRAMMAR of the ITALIAN LANGUAGE, with Exercises. By PIETRO BACHI. Instructer at Harvard University. 12mo.

This book has been compiled from the best Grammars of the Italian tongue, extant in Italian, English, and French, special use having been made of Barberi's celebrated "Grammaire des Grammaires Italiennes." It is divided into four parts, Pronunciation, Orthography, Analogy, and Syntax. The pronunciation is more fully treated than in any other Grammar, and illustrated by English combinations of letters representing the Italian sounds. Every Italian word is accented throughout the book, so that the pupil grows familiar with the genius of the language in this respect, while he is acquiring a knowledge of its grammatical structure. The verbs are given with unusual fulness, and the irregular verbs are arranged and exhibited after a new method very convenient for reference.— The rules of the Syntax are supported by examples carefully cited from the best writers, and followed by appropriate exercises. A copious Alphabetical Index closes the volume.

In press, and will soon be published. JUVENAL'S SATIRES, with English Notes. An expurgated edition of the Satires of Juvenal, with copious English Notes, after the plan of Gould's edition of Virgil. By the editor of Greek Delectus, Phædrus, &c.

IRVING'S ELEMENTS of ENGLISH COMPOSITION; serving as a sequel to the study of Grammar. By DAVID IRVING, LL. D. Author of the Lives of the Scottish Poets. Second American from the sixth London edition. Price $1,25.

Extracts from the Preface.

Though it was my principal object to treat of prose composition, yet a few observations on poetry incidentally occur. The remarks which have been suggested with regard to the nature of figurative language, apply equally to prose and to poetry; but the poets have furnished me with the most copious and beautiful illustrations.

The rules of criticism are more successfully inculcated by particular examples than by general precepts. I have therefore endeavored to collect abundance of apposite quotations, in order to illustrate every branch of the subject.

An INTRODUCTION to LINEAR DRAWING; translated from the French of M. FRANCŒUR; with alterations and additions to adapt it to the use of Schools in the United States. To which is added, the Elements of Linear Perspective; and Questions on the whole. By WILLIAM B. FOWLE, Instructer of the Monitorial School, Boston. Price 75 cents.

From the Translator's Preface.

An elementary treatise on Drawing, adapted to the use of common schools, cannot but be well received. Besides the professions which make the art of drawing their particular study, anatomists, naturalists, mechanics, travellers, and indeed all persons of taste and genius, have need of it, to enable them to express their ideas with precision, and make them intelligible to others.

Notwithstanding the great utility of this branch of education, it is a lamentable fact, that it is seldom or never taught in the public schools, although a very large proportion of our children have no other education than these schools afford. Even in the private schools where drawing is taught, it is too generally the case that no regard is paid to the geometrical principles on which the art depends. The translator appeals to experience when he asserts, that not one in fifty of those who have gone through a course of instruction in drawing, can do more than copy such drawings as are placed before them. Being ignorant of the certain rules of the art, (and they are the most certain because mathematical,) they are always in leading strings, and unless endowed with uncommon genius, never originate any design, and rarely attempt to draw from nature. It is to remedy this defective mode of teaching, that the translator has been induced to present this little work, on the elements of drawing, to the American public.

Preface to the Second Edition.

The favorable reception of the first edition of this Treatise, has induced the Translator to revise it carefully, and to add to it a Second Part, containing the elements of *Perspective Drawing*, to which the First Part is a good introduction.

Questions, also, upon the more important parts of the book are added ; and the Translator hopes that this more correct and enlarged edition will meet with the same favor that a liberal public has bestowed upon its predecessor.

The LATIN READER. Part First. From the fifth German edition, by FREDERIC JACOBS, Editor of the Greek Anthology, the Greek Reader, &c. &c. Edited by GEORGE BANCROFT. Stereotype edition. Price 87½ cents.

The Latin Reader, which is here published, was compiled by Professor FREDERIC JACOBS, of Gotha, who having long been engaged in the cares of instruction and the pursuits of a scholar, is in every respect qualified to make judicious selections for the purposes of teaching.

The editor, in publishing this work in America, has been influenced by a sincere belief, that it forms an easy introduction to the language and character of the Roman world. His duties as a teacher led him to the comparison of many similar works now used in England and on the continent. This seemed to him the best ; and having already used it in the school with which he is connected, he has found his opinion confirmed by his experience.

This work is very fast taking the place of Liber Primus, Historiæ Sacræ, Viri Romæ, &c.

The advantages of this work are, that it proceeds by gradual and easy steps, from the examples of the first principles of Grammar, usual in Primary books, to the more difficult Latin of the authors to be studied next in course—thus including in *one* volume what commonly occupies two or more. The necessity of adapting the matter to the gradual progress of the pupil has secured a variety of selection, sufficient to keep the attention excited ; and thus to obviate the motive for a frequent *change* of works : while the amount of useful knowledge bound up in these pages exceeds that to be found in most other books of the same description.

In the *stereotype* edition, the Dictionary has been new modelled entire. It is of a size commensurate with the text, for which it is designed ; and may, for some time, at least, stand the pupil instead of the more cumbrous and expensive works of Entick and Ainsworth—with the additional advantage of being based upon Adam's Latin Grammar ; and having the minute irregularities of Declension, Gender, and Tense, and also the component parts of compound words, noted in full, by a gentleman of well known accuracy and judgment.

The LATIN READER. Part Second. Chiefly from the fourth German edition of F. JACOBS, and F. W. DÖRING. Edited by GEORGE BANCROFT. Stereotype edition. Price 75 cents.

The present continuation of the Latin Reader has for its object, to provide a work suited to the purposes of instruction in the Latin language, of a classical character, interesting to the young mind, and conveying useful information. The first part of this little volume contains select fables from Phædrus; these are followed by extracts taken almost entirely from Cicero and Livy ; the volume closes with an abridgment of Justin, for the excellence of which the name of Jacobs is alone a sufficient recommendation. Most of the "Short Narrations" were selected by Döring, who acted in concert with Jacobs. A few more have been added from the twelfth German edition of an elementary work, compiled by Bröder. In presenting to the public this edition, it is only necessary to say, *that the text* has undergone a thorough revision ; that uniformity has been introduced in the orthography, which is now made to conform to that of the Latin

Dictionaries in common use; and that equivocal words have been marked with accents in the usual way, and speeches distinguished from the narrative by inverted commas.

The LATIN TUTOR, or an Introduction to the making of Latin, containing a copious exemplification of the rules of the Latin Syntax from the best Authorities. Also rules for adapting the English to the Latin idiom. The use of the particles exemplified in English sentences designed to be translated into Latin. With rules for the position of words in Latin composition. Price 87½ cents.

The object of this work is to furnish the Latin student with a series of exercises adapted to familiarize to his mind the inflexions of words, and the application of the rules to syntax, and to lead him to such a knowledge of the structure of the language as may enable him to read and write it with ease and propriety.

The materials of which it is composed have been drawn from the purest sources, and will be found to possess intrinsic merit in sentiment, clothed in a rich variety of elegant and classical expression, the order and arrangement, it is hoped, will be found correct and perspicuous.

But the principle point on which the claims of this work are rested, is, that it endeavors to present, in every part, a genuine Latin style, in place of that nondescript style, produced by conforming the Latin words to the English collocation, which occupies a considerable proportion of every work on this subject which has fallen within our knowledge.

This work is now used in the Boston Latin School.

LETTERS on the GOSPELS. By Miss HANNAH ADAMS. Second edition. Price 75 cents.

Extract from the Christian Examiner and Theological Review.

We have been very highly gratified by the perusal of this little book, which, coming out with all the modesty, simplicity, and real learning, which distinguish its author, is calculated we think, to be of very important service in the cause of true religion.

We have rarely seen so much valuable knowledge brought in so small a compass, or in so attractive a manner, to the level of youthful minds.

An ABRIDGMENT of MURRAY'S ENGLISH GRAMMAR. Containing also Punctuation, the Notes under rules in Syntax, and Lessons in Parsing. To the latter of which are prefixed, Specimens illustrative of that Exercise, and false Syntax to be corrected. All appropriately arranged. To all which is adapted, a New System of Questions. From the second Portsmouth edition, enlarged and improved. By SAMUEL PUTNAM. *Stereotype edition.* Price 19 cents.

The sale of the former editions of this work has encouraged the editor to offer the public another, containing, as he would hope, some valuable improvements.

The object of the questions interspersed through this Grammar, is to lead the learner, while committing his lesson, to discover its meaning and application.

As many scholars never use any other than the cheap editions of the Grammar, it is certainly important that such an abridgment should contain, if possible, all the necessary rudiments.

There is, among some, a disposition to abandon Murray entirely, and adopt divisions and arrangements altogether new. Every new system will, without doubt, present some new and valuable views. But whether, upon the whole, any single system can at present be found, more scientific, or affording greater facilities in learning to speak and write our language correctly, is greatly to be doubted.

To the Editor.

SIR,—I have examined, with some attention, the third edition of an Abridgment of Murray's English Grammar, published by HILLIARD, GRAY, & Co. Having, for a considerable time, used the former editions of the same work, I was, in some measure, prepared to appreciate this.. The lessons in parsing are well chosen and the arrangement of them a valuable improvement. The NEW SYSTEM of QUESTIONS has long been a *desideratum* in an introduction to the English Grammar, and seems perfectly to answer the end designed. In short, I regard this little book as a highly valuable acquisition to our schools; far preferable to any work of the kind that has come under my observation, and am persuaded that your labor in this department of early science will meet all the encouragement you can desire.

I am, Sir, &c.   ORANGE CLARK, *Principal*
Portsmouth, June 13th, 1827.   *of the Portsmouth Lyceum.*

MURRAY'S INTRODUCTION to the English Reader, or a Selection of Pieces in Prose and Poetry, calculated to improve the Younger Classes in Reading, and to imbue their minds with the love of virtue. To which is added, Rules and Observations for assisting children to read with Propriety. Improved by the addition of a Synonymising Vocabulary, of the most important Words, placed over the sections, from which they are selected, and divided, defined, and pronounced according to the principles of JOHN WALKER. Walker's Pronouncing Key, which governs the Vocabulary, is prefixed to the work.   Price 37½ cents.

"This Introduction is full of simple, natural, and interesting pieces. It is we think the best juvenile selection in the English language. It produces moreover a fine animation and an intelligent style of reading, which are great aids to general improvement. The present edition of this useful work, has an important addition to recommend it, as mentioned in the title given above."—*Journal of Education, Vol.* 2, *No.* 9.

The Introduction to the English Reader is considered the best of Mr. MURRAY'S reading books; and this is no small praise, when the popularity and excellence of them all is considered.

The Introduction now offered to the public is improved by the addition of a vocabulary of the most important and difficult words prefixed to each section, showing their pronunciation and definition. This mode is far preferable to having a general vocabulary appended to the work, which is troublesome to the scholar and therefore apt to be neglected, or, if much recurred to, causes the book to be sooner defaced and destroyed;—and much better than having the pronunciation given in the body of the work, which blurs and disfigures the page and renders the book, to the young, almost illegible.

NEUMAN and BARRETTI'S DICTIONARY of the Spanish and English Languages; wherein the words are correctly explained, agreeably to their different meanings, and a great variety of terms, relating to the Arts, Sciences, Manufactures, Merchandise, and Trade, elucidated. Stereotype edition, carefully revised, and enlarged by the addition of many thousand words extracted from the writings of the most Classical Spanish and English Authors, many of which are not to be found in any other Dictionary of those Languages; and also great additions from the Dictionaries of CONNELLY and HIGGENS, the Spanish Academy, &c. To which are added Directions for finding the difference between the Ancient and Modern Orthography, by F. SALES, Instructer of French and Spanish at Harvard University, Cambridge.   2 vols. 8vo.

An INTRODUCTION to SYSTEMATIC and PHYSIOLO-GICAL BOTANY. By THOMAS NUTTALL, A. M., F. L. S., &c., Lecturer on Botany and Zoology, and Curator of the Botanic Garden connected with Harvard University, Cambridge.   1 vol.   Price $2,00.

The present work forms a happy exception to those Introductory Treatises upon different subjects, which are the offspring of avarice, or of the pride of authorship.

The work is accompanied by twelve very beautiful lithographic engravings; and its entire execution is characterized by neatness and precision.

In conclusion, we would only remark, that it has fully answered the expectations we had formed of it, from a knowledge of the high attainments of its author, and that, in our opinion, it constitutes by far the most valuable treatise that can be put into the hands of a person just commencing this delightful study. To those who are acquainted with Mr. NUTTALL'S former productions, it need not be mentioned, that his style is simple, condensed, and highly perspicuous; precisely what a style ought to be in all works of a similar nature.—*American Journal of Science and Arts.*

The NATIONAL READER; a Selection of Exercises in Reading and Speaking, designed to fill the same place in the Schools of the United States, that is held in those of Great Britain by the compilations of Murray, Scott, Enfield, Mylius, Thompson, Erving, and others. By

John Pierpont, Compiler of the American First Class Book. 1 vol. Price 75 cents.

"Induced by esteem for the compiler, as well as by a deep interest for whatever concerns the subject of education, we have examined the National Reader with care, and with satisfaction."

"Finding the work thus deserving of favor, we earnestly recommend it to the adoption of all teachers of youth who desire to instil into their pupils a taste for moral and literary beauty, and a love of country."—*National Intelligencer*, Oct. 11, 1827.

"The National Reader is designed for the *common schools* of the United States, but it will be found a valuable introduction to the First Class Book in those higher institutions in which that work is used. After carefully and thoroughly examining this compilation, we confidently and with pleasure recommend it to the notice of teachers, school committees, and all others interested in the education of the young. The selection of lessons is peculiarly rich, and sufficiently diversified."— "We could not easily name a book of equal size which contains so great a variety of classically chaste and interesting matter; and we think it well worthy of a place in every parlor as a volume of elegant extracts."—*American Journal of Education*, Oct. 1827.

"This is one more added to the many fine selections for schools which have been published within a few years."—"The National Reader contains a suitable proportion of extracts from our own writers, both in poetry and prose, and is manly enough not to think it anti-national to borrow from the stores of England. We do not know that a better book of the kind could be made."—*Christian Examiner*, Sept. and Oct. 1827.

"We are anxious to add our feeble testimony to the excellence of this compilation. It is a volume calculated to fill a high and an important place in our schools. The selections are made chiefly from American writers of high standing, and they are happily adapted to convey useful information, to improve the taste, to interest the feelings, and to leave the best moral and religious impressions. We were exceedingly pleased to observe its serious character; and we should think that no one could attentively peruse it, without being the wiser and better; without being more sensible of his obligations to be virtuous and devout; without a deeper conviction that he is immortal and responsible. We are fully persuaded that, where it is used in schools, it cannot fail to produce the most desirable effect on the dispositions and conduct of the youth."—"In preparing this volume the compiler has conferred a favor on the community, which we trust they will cheerfully acknowledge by extensively introducing it into their schools."—*Greenfield Gazette and Franklin Herald*, Dec. 1827.

Our limits do not permit us even to make extracts from the highly favorable notices which have been taken of the National Reader by the *Christian Intelligencer and Eastern Chronicle* of Maine, the *Courier*, *Statesman*, and *Galaxy* of Boston, the *Troy Sentinel*, the *Western Monthly Review, for* Oct. 1827, and the *Statesman, Daily Advertiser, Albion, Morning Courier, Mirror*, and other journals of New York.

In Press. The AMERICAN SCHREVELIUS, or Greek and English Lexicon, new and improved edition.

The basis of the work is Schrevelius's well known Lexicon; which, on the whole, in the present state of Greek studies in this country, was thought preferable to any other manual adapted to *the use of schools*. Schrevelius's work was more particularly intended for the Old and New Testaments, Homer, Hesiod, Musæus, Theognis, Pythagoras, and other Gnomic Authors, Isocrates, Æsop, &c.; the author also made use of Portus's Ionic and Doric Lexicons and the Lexicon to Pindar and the other Lyric poets. It was published several times on the continent of Europe during the author's life; and within that period was also republished in England by Hill, who enlarged it considerably, more particularly with words from the New Testament, the Septuagint, and the principal poets and orators, as well as the school books of the day. Besides the editorial labor bestowed upon it in England, it has received improvements in France, where a valuable edition of it was published in 1779 by the celebrated scholar Vauvilliers. Of the other editions, we have before us the Italian one in folio, and a German one, reprinted from the Paris copy, at Vienna in 1822, under the editorial superintendence of Kritsch; who justly observes, that the Lexicon, as now published, is very different from the ancient editions both in copiousness and explanations; and, in its present state it may with propriety be recommended to the student in Greek literature. The significations given in this work are more copious than the Latin ones of Schrevelius. It has been the intention of the editors, that the work should comprehend all the words which are to be found in Professor Dalzel's Collectanea Majora and Minora, Jacob's Greek Reader, and the other books now studied in our schools and other seminaries of learning.

The improvements made upon the common Schrevelius, in the present edition will amount to not less than *ten thousand new* articles and very numerous additions to the original articles of the work.

The explanations of the uses of the *prepositions* and *article*, which were the subject of particular attention in the former edition, have been still further improved in the present one. Another improvement (and one which was not adopted in any edition of Schrevelius till after this work was begun) is the marking of the *quantities* of the doubtful vowels; and in the present edition this has been more minutely attended to than in the former.

During the progress of the work almost all the Greek Lexicons extant have been occasionally consulted. Those which have been most constantly resorted to are—Schneider's admirable Griechisch-Deutsches Wörterbuch (or Greek and German Dictionary) and the Greek and German Lexicon of Riemer, who has added much new and valuable matter to Schneider's labors; Planche's excellent Dictionnaire Grec-François,—Donnegan's New Greek and English Lexicon, the basis of which is Schneider's;—Jones's Greek and English Lexicon, and the improved edition of Hedericus; and, for the Scripture words, Schleusner's well known Lexicon and Wahl's Greek and English Lexicon by Mr. Robinson. Besides these aids, as much use, as was practicable in a work of this size, has been made of the labors of eminent critics and commentators on the Greek Classics.

PHÆDRI FABULÆ EXPURGATÆ. Accedunt Tractatus de versu Iambico, Notulæ Anglicæ et Questiones. In usum Scholæ Bostoniensis. Price 62½ cents.

Phædrus comes strongly recommended to the study of the young. He has clothed the most beautiful of Æsop's fables in an elegant, pure, and simple style. With the interest of the Fable he has united the graces of poetry, and the graver wisdom of philosophy. His writings are eminently pleasing and manly ; and he has succeeded beyond most authors in what he states to be his design, at once to *amuse* and to *instruct*. In the present edition, the text has been carefully corrected, according to the best texts, the Notes are very full, containing a good deal of Mythological and Historical matter together with Questions annexed. But the most important improvement is the attention devoted to *Iambic Verse*. This measure, a favorite in the English as well as in the Latin and Greek languages, has been hitherto strangely neglected in our schools. It was thought, that in publishing an edition of Phædrus, one of the best and easiest specimens of Iambic writers, a favorable opportunity was offered for introducing the study of this elegant and popular measure, and of course an insight into all the Latin and Greek Tragedy and Comedy. Accordingly some few, but comprehensive remarks on Iambics have been subjoined including a scale of feet adapted to Phædrus ; and the Notes have been interspersed with explanations of all the difficult lines. The work is prepared and printed according to the plan of Gould's Virgil and Ovid, particularly for the use of the Boston Public Latin School.

The RATIONAL GUIDE to Reading and Orthography ; being an attempt to improve the arrangement of Words in English Spelling Books, and to adapt the Reading Lessons to the comprehension of those for whom they are intended. By WILLIAM B. FOWLE, Instructer of the Monitorial School, Boston. Price 25 cents.

'The matter and the arrangement of this little volume possesses much originality ; both are happily adapted to the capacity of young children, and are excellently suited to aid a gradual and sure progress in the principles of reading.

The reading lessons which are interspersed with the columns, are simple and intelligible ; they are all written in a very interesting style ; and many of them convey useful moral instruction.

From a pretty extensive acquaintance with similar school books issued from the English press, we are enabled to make a comparison which is highly favorable to Mr. FOWLE's. There is no work of the kind, as far as we know, which is equally well adapted to the use of beginners in reading and spelling ; or which an *instructer* may use with so much advantage and pleasure.'—*Journal of Education*.

This Spelling Book is elegantly stereotyped, and although, by its peculiar arrangement and classification of words and lessons, it is admirably adapted for the use of beginners, still it contains as many words, with less extraneous matter than any other Spelling Book, and is suitable for the higher classes also.

A RHETORICAL GRAMMAR; in which the common improprieties in Reading and Speaking are detected, and the sources of Elegant Pronunciation are pointed out. With a Complete Analysis of the Voice, showing its specific·Modifications, and how they may be applied to different species of sentences, and the several figures of Rhetoric. To which are added Outlines of Composition, or plain Rules for writing Orations and Speaking them in Public. By JOHN WALKER, Author of the Critical Pronouncing Dictionary, Elements of Elocution, &c. Second American edition. 1 vol. 8vo. Price $2,25.

Extract from the Preface.

The want of rules for composition, so essential in rhetoric, has been supplied in this edition from the best source—BLAIR's Lectures: and what was deficient even in these has been furnished from Prof. WARD's Lectures on Oratory:—so that with the original matter on the elegant pronunciation of words, on accent, emphasis, and inflexion of voice, and the proper pronunciation of the figures of rhetoric, it is presumed the present work is the most perfect of its kind in the language.

The NEW TESTAMENT of our Lord and Saviour Jesus Christ; in which the Text of the Common Version is divided into paragraphs, the punctuation in many cases altered, and some words, not in the original expunged. 12mo. Price 50 cents. (Called the Revised Testament.)

It is well known to the learned, and should be to all, that the division of the Bible into chapters and verses, the punctuation, and the words usually printed in italics, are of no authority whatever. The several Books of the New Testament, according to the best authority we can get, were written in an uniform character, without capitals, without chapters, without verses, without punctuation, or any break or other index by which to determine whether a particular letter belonged to this or that word, or whether a particular word belonged to this or that sentence; the sense was the only guide to the proper division.

The basis of the divisions which are found in this edition is the Greek Testament edited by Knapp.

The PHILOSOPHY of NATURAL HISTORY, by WILLIAM SMELLIE, Member of the Antiquarian and Royal Societies of Edinburgh. With an Introduction and various Additions and Alterations, intended to adapt it to the present state of knowledge; by JOHN WARE, M. D. Fellow of the Massachusetts Medical Society, and of the American Academy of Arts and Sciences. Second edition. Price $2,25.

SMELLIE's Philosophy of Natural History has been for many years one of the most popular, as it is one of the most instructive treatises upon the subjects to which it relates. Although it does not profess to be a complete system of Natural History it still contains a great variety of facts interesting to the scientific Naturalist; and although not adapted to all his wants, is at least calculated to minister to his pleasure. It is more particularly intended for the use of those who are lovers of nature in general; who admire and love to study her works, but have not leisure or ability to acquaint themselves with the technical details of science. It dwells principally upon such subjects as are within the comprehension of all—such as relate to the manners characters and habits of animals—the skill displayed in the construction of their habitations—the mode in which they obtain their food—their means and instruments of defence and attack—their care of and attention to their offspring—their docility, powers of imitation, &c. &c. Hence the work has been always found one of the most interesting which could be put into the hands of the young.

In the American edition no essential alterations have been made in its plan or details. An introduction has been prefixed in order to possess the reader of a few of the elementary principles of the science of Natural History as established at the present day. In the body of the work a few additions have been made, particularly to the chapters on Physiology, and a few chapters have been omitted, as relating to subjects which would not be interesting or useful to those for whom it was chiefly intended. It is believed, that in its present form it is calculated to interest the minds of the young, to convey to them much useful knowledge, and to give them a taste for the wonders and beauties of nature.

3

The THIRD CLASS BOOK; comprising. Reading Lessons for young scholars. Second edition. Price 37½ cents.

The principal end in view, in the compilation of this work, has been to present to the younger classes in our common schools, a book not only adapted to their intellectual capacities, but also adapted to their sympathies and feelings. A story may be perfectly intelligible and powerfully interesting to a child; and at the same time the interest it creates, and the feelings it excites, are those which should belong only to maturer years. For a reading lesson, a story should possess just interest enough to engage and fix the attention of the reader;—any thing short of, or beyond this effect, is aside from the purpose.
The second edition is enlarged, and greatly improved.

The ART of READING : or Rules for the attainment of just and correct enunciation of written language. Mostly selected from WALKER's Elements of Elocution, and adapted to the use of Schools. Price 50 cents.

WALKER's Elements of Elocution is a work, which has enjoyed a great reputation both in England and in this country. The correctness of its principles is generally admitted, and the rules it contains are allowed to be the most accurate guide we possess on the subject of reading. It is not however, in its present form, well suited to the purposes of a school book. The volume is too expensive for general use, and contains much which can neither be applied nor understood by the majority of pupils. It occurred to the compiler, that an abridgment of Walker's Treatise divested of all minute disquisition, and rendered strictly practical in its character, might be a useful manual for schools. This idea has been confirmed by some intelligent friends whom he has consulted, and hence this little volume is now offered to the public. The alterations which have been attempted in the present compendium, are not numerous. They consist in occasionally varying the order of the rules, in simplifying their language, and in supplying a few additional examples for practice. Where these changes occur, it is hoped their utility will appear sufficient to justify their introduction.

Will speedily be Published. TACITI OPERA OMNIA, Notis Anglicis Illustrata. 2 vols. 12mo.

This work will be executed for the use of schools on the plan of Gould's editions of Virgil, Ovid, and Horace.

The ELEMENTS of GREEK GRAMMAR, with Notes by R. VALPY, D. D., F. A. S. Sixth Boston edition—carefully revised and corrected at the University Press, Cambridge. Price $1,00.

WILKINS'S ELEMENTS of ASTRONOMY, illustrated with Plates, for the use of Schools and Academies, with Questions. By JOHN H. WILKINS, A. M. Fifth edition. Stereotyped. Price 87½ cents.

The design of this work is to exhibit the leading facts and to illustrate the leading principles of Astronomy in a manner interesting and useful to those scholars who do not intend to pursue the subject to great extent. It may be studied without a knowledge of the higher branches of mathematics; and contains familiar illustrations of the most striking phenomena of nature. The work has passed through four editions and the fifth is now published. It is used in the principal schools in Boston and the vicinity; and is coming into very general use.
This work is peculiarly adapted for the use of Academies and the higher classes in common Schools.

The BIBLE CLASS TEXT BOOK; or Biblical Catechism, containing questions historical, doctrinal, practical and experimental. Designed to promote an intimate acquaintance with the inspired volume. By HERVEY WILBUR, A. M. Seventeenth edition, revised, improved, enlarged, and stereotyped, with practical questions annexed to the answers. Price 37½ cents.

A CATECHISM in three parts *Part first*, containing the elements of Religion and Morality; designed for children. *Part second*, consisting of Questions and Answers, chiefly historical, on the Old Testament. *Part third*, consisting of similar Questions and Answers on the New Testament; designed for children and young persons. Compiled and recommended by the *Ministers of the Worcester Association in Massachusetts.* Third edition. Price 12½ cents.

The foregoing little book, in part compiled from various sources, and in part original, it is believed, will be found to contain important religious instruction in a form adapted to the minds of children and young persons; and it is accordingly recommended by the ministers of the Worcester Association to the people of their respective charges, and to all parents and instructers of youth, with the hope and prayer that, by the blessing of God, it may be instrumental in promoting a knowledge of the holy Scriptures, and in forming youthful minds to virtue and piety.

A PRIMER of the ENGLISH LANGUAGE for the use of Families and Schools. By SAMUEL WORCESTER. Stereotype edition. Price 12½ cents.

"The author of this invaluable little manual, has done more for the health and the happiness, as well as the intellectual improvement of infant learners, and for the relief and aid of mothers, than if he had published an octavo volume on the philosophy of instruction or on the duties of mothers. Instead of formally discussing his method, (which appears to us decidedly the best ever offered to the public,) we will furnish some extracts from the work itself, which will serve to give an idea of the spirit of the plan.

'In order to teach this Primer, it will be absolutely essential that the instructer should abandon the common method of teaching children to read and spell The author, therefore, earnestly requests all teachers to attend carefully to the directions which precede the several classes of lessons.

'It is not, perhaps, very important that a child should know the letters before it begins to read, it may first learn to read words by seeing them, hearing them pronounced, and having their meaning illustrated; and afterwards it may learn to analyse them, or name the letters of which they are composed. Those instructers who choose to adopt this method, may commence with Lesson I, and require the scholar to read all the words in six or eight lessons, without attempting to spell them; and then to recommence the book with the alphabet, and spell the words selected from each reading lesson.'

[Further directions will be found in the Author's Preface.]

"A work like this, which so happily addresses itself to the very elements of infantine thought and feeling, it would be idle to praise. The book carries, within itself, its best recommendation to a parent's heart."—*Journal of Education.*

Worcester's Elementary Works on History and Geography.

ELEMENTS of HISTORY, Ancient and Modern, with Historical Charts. By J. E. WORCESTER, A. A. S., S. H. S. Third edition. Price of the *History* $1,00. Price of the *Historical Atlas* $1,50.

The Historical Atlas which accompanies the volume, comprises a series of *Charts*, formed on a new plan, and affording means of facilitating the study of *History* similar to what are afforded by *Maps* in the study of *Geography*. It contains the following *Charts:*—

1. Chart of General History.	6. Historical Chart of England.
2. Chart of Sacred History.	7. Historical Chart of France.
3. Chart of Ancient Chronology.	8. Chart of American History.
4. Chart of Sovereigns of Europe.	9. Chart of Biography.
5. Chart of Modern Chronology.	10. Chart of Mythology.

Extract from the Advertisement to the Third Edition.

"The approbation which has been expressed of the plan and execution of this work, by the different literary journals, and by various respectable instructers, and distinguished literary gentlemen, and likewise by the public, as indicated by the rapid sale of the first and second editions, call for the grateful acknowledgments of the Author.

"In this third edition the volume has been carefully revised, and the quantity of matter increased. A Chart of *Sacred History* has been added to the Atlas, and all the other Charts have been altered and improved."

This work has been highly recommended by various literary gentlemen.—*President Kirkland* and *Professors* Hedge and *Willard*, of Harvard University, and the *Rev. Dr. Beasley*, Provost of Pennsylvania University, say in their recommendation;—" We can cheerfully recommend it as the best elementary work of the kind with which we are acquainted."—*President Lindsley*, of Cumberland College, says;—" I give it the decided preference to every work of the kind with which I am acquainted."

QUESTIONS adapted to the use of WORCESTER'S Elements of History. Price 18¾ cents.

WORCESTER'S EPITOME of HISTORY, with Historical and Chronological Charts. Price of the *Book* 50 cents. *Atlas* 50 cents.

The Atlas contains the four following Charts:—

1. Chart of General History.   3. Chart of Modern Chronology.
2. Chart of Ancient Chronology.   4. Chart of American History.

" This *Epitome of History* and the corresponding *Epitome of Geography* by the same author, are well adapted, as regards size and cost, to the circumstances of primary and common schools; and the characteristic care and exactness of the author leave little to desire in regard to the character of these books, as works adapted to the objects of education. We earnestly hope that school committees will examine this work and give it a place among their school books."—*Journal of Education.*

WORCESTER'S ELEMENTS of GEOGRAPHY, Ancient and Modern; with an Atlas, Ancient and Modern. Price of Geography 87½ cents. Price of *Modern Atlas* $1,00. Price of *Ancient Atlas* 87½ cents.

Maps in the Ancient Atlas.

1. Roman Empire.   4. Asia Minor.
2. Italy.   5. Palestine.
3. Greece.

Maps in the Modern Atlas.

1. The World.
2. North America.
3. United States.
4. Eastern and Middle States.
5. South America.
6. Europe.
7. France, Germany, Netherlands, Switzerland, &c.
8. England, Scotland, and Ireland.
9. Asia.
10. Africa.

In this Geography scholars are examined for admission into Harvard and other Colleges. It is also, by order of the School Committee, used in all the public Grammar Schools in Boston, and to great extent through the country.

" Mr. Worcester's Geography appears to us a most excellent manual. It is concise, well arranged, free from redundancies and repetitions, and contains exactly what it should, a brief outline of the natural and political characteristics of each country. The tabular views are of great value."—*North American Review.*

" I cannot hesitate to pronounce it, on the whole, the best compend of Geography for the use of Academies, that I have ever seen."—*Rev. Dr. S. Miller, of Princeton.*

" Of all the elementary treatises on the subject which have been published, I have seen none with which I am on the whole so well pleased, and which I can so cheerfully recommend to the public."—*President Tyler, of Dartmouth College.*

WORCESTER'S EPITOME of GEOGRAPHY; with an Atlas. Price of *Epitome* 50 cents. Price of *Atlas* 75 cents.

The Atlas contains the following Maps:—

1. Map of the World.
2. North America.
3. United States.
4. South America.
5. Europe.
6. British Isles.
7. Asia.
8. Africa.
9. Roman Empire.
10. Palestine.
11. Comparative Heights of Mountains.
12. Comparative Lengths of Rivers.
13. Statistical Summary of the United States.
14. Statistical Summary of Europe.
15. Statistical Summary of the Globe.

" Mr. Worcester's success as a geographer renders it unnecessary for us to say any thing more of this little work, than that it bears all the characteristic marks of his former productions. He is accurate, clear, and remarkably happy in condensing the most important particulars, and bringing them down to the ready apprehension of children. The author has one merit almost peculiar to himself.

He has taken unwearied pains to designate the accurate pronunciation of proper names. This is of great utility."—*North American Review.*

"It would be difficult to collect within the same limits so much exact, useful, and well chosen matter, as is contained in this little volume, and the accompanying Atlas. Mr. Worcester's books have all been distinguished for accuracy and clearness. This is no less so than those which have preceded it—We have seen no book so well adapted to the wants of teachers and learners."—*Christian Examiner.*

SKETCHES of the EARTH and its INHABITANTS; comprising a Description of the Grand Features of Nature; the Principal Mountains, Rivers, Cataracts, and other Interesting Objects and Natural Curiosities: also of the Chief Cities and Remarkable Edifices and Ruins, together with a View of the Manners and Customs of Different Nations: *Illustrated by One Hundred Engravings.* 2 vols. Price $3,50.

Extracts from Reviews, &c.

"We have attentively perused these 'Sketches,' and have no hesitation in saying that we know of no similar work, in which instruction and amusement are so much combined. The accuracy of the statements, the brevity and clearness of the descriptions, the apposite and often beautiful quotations from books of travels and from other works, continually excite and gratify the curiosity of the reader."— *Christian Spectator.*

"We consider the 'Sketches' well suited to give a large fund of entertainment and instruction to the youthful mind."—*North American Review.*

"We know of no book which would be more suitable to be read by scholars in our higher schools, and which would excite more interest in the family circle."— *R. I. American.*

"These volumes are extremely entertaining, and may be recommended to the perusal of those even, who conceive themselves to be past the necessity of elementary instruction."—*Christian Examiner.*

"The 'Sketches,' &c. form a most valuable companion to the '*Elements of Geography,*' admirably calculated to interest the attention, and impart useful knowledge to our youth."—*Roberts Vaux, Esq.*

"The work is, in my opinion, ably executed, and well fitted to be both popular and useful."—*Rev. Dr. S. Miller.*

WORCESTER'S GEOGRAPHICAL DICTIONARY, or Universal Gazetteer, Ancient and Modern. Second edition, in 2 vols. Bound. Price $11,00.

A GAZETTEER of the UNITED STATES abstracted from the Universal Gazetteer of the Author. With enlargement of the principal articles, by J. E. WORCESTER. 1 vol. 8vo. Bound, price $2,50.

Will speedily be Published. XENAPHONTIS ANABASIS, cum Notulis Anglicis et Quæstionibus. 2 vols. 12 mo.

This work will be prepared for the use of Schools, on the same plan with Gould's edition of Virgil, Horace, and Ovid.

The FRENCH ACCIDENCE or Elements of French Grammar. By WILLIAM B. FOWLE, Instructer of the Monitorial School, Boston.

The prevailing custom of requiring young children to commence the study of the French language, has led to the publication of this Accidence. The author found that his younger pupils were unable distinctly to collect the leading principles of French Grammar from the bulky works commonly used, and that the more advanced pupils often experienced some difficulty in finding the information they sought, not because it was not in the book, but because there was so much else there. It is believed that all the essentials of French Grammar are contained in this compend, and that they are so arranged as to be distinct and obvious. Should this Accidence be found useful, it will be followed by a small volume of Exercises adapted to it.

GARDNER'S TWELVE INCH GLOBES.
$26,00 per pair.

These Globes are now very generally used in the Schools and Academies of New England.

RECOMMENDATIONS.

Cambridge, Jan. 23, 1824.

Mr. J. W. GARDNER appears to have made himself acquainted with the best methods of constructing artificial Globes, and to have used all due care in the construction of his work. The stars are laid down singly, and the number and outline of the constellations are given, according to the latest and most approved catalogues and charts of the heavens. The terrestial Globe is understood to have been constructed in a similar manner, by means of the best tables of the latitude and longitude of places. These Globes are accordingly recommended as well adapted to the purposes of elementary instruction in Geography and Astronomy.

JOHN FARRAR,
Prof. of Math. and Nat. Phil. in Harvard University.

Boston, Jan. 22, 1824.

Rev. and Dear Sir,—I have examined, with a great deal of care, Mr. Gardner's Terrestrial Globe, and find it very accurately executed according to the latest discoveries and the best tables. The coasts are carefully laid down, with an extraordinary degree of minuteness, from the best established authorities. The outlines are consequently very exact. All important places in the interior of continents, where the latitude and longitude have been well ascertained, are also laid down with the same care.

With great respect, your obedient servant,   GEO. B. EMERSON.
Rev. President KIRKLAND.

Harvard University, Cambridge, 23 Jan. 1824.

Sir,—After the opinion expressed by Professor Farrar of your ability and fidelity in the construction of your Globes; and after the testimonial of Mr. George B. Emerson, who has examined your Terrestrial, and of Mr. James Hayward, who has paid particular attention to your Celestial Globe, (one of these gentlemen having been lately and the other being now in the department of Mathematics, Natural Philosophy, and Astronomy in this University, and both well known for their accurate acquaintance with these branches of science,) I can have no doubt that these productions of your skill and labor are fully entitled to peculiar favor and patronage.   I am, Sir,.with esteem, your obedient servant,

J. T. KIRKLAND, *President.*
Mr. J. W. GARDNER.

GARDNER'S NEW FOUR INCH GLOBES.
Price $2,00 each.

In consequence of the increased demand for GARDNER's twelve inch GLOBES, the author has been induced to publish a pair of four inch Celestial and Terrestial. corresponding in every respect with the twelve inch, except the omission of the names; which present an entirely new and interesting exercise in Geography and Astronomy.

BOSTON BOOKSTORE.

HILLIARD, GRAY, & Co. Publishers, Booksellers

and Stationers, No. 134, Washington street, Boston, keep constantly for sale a great variety of English, French, Italian, Spanish and German Books, in the various branches of Learning. An extensive stock of the most valuable LAW BOOKS.

ALSO,—A supply of the finest LONDON STATIONARY. Orders for Books of every description executed with care and promptness, and on the most moderate terms.

☞ Social Libraries and Literary Institutions, supplied with miscellaneous books at a great discount. Old and rare Books often on hand for sale at very low prices.

T. R. MARVIN, PRINTER, 32, CONGRESS-STREET, BOSTON.

ridge, 23 Jan. 1824.
r ability and fidelity
1 of Mr. George B.
mes Hayward, who
of these gentlemen
ent of Mathematics,
both well known for
I can have no doubt
tled to peculiar favor
lient servant,
AND, *President*.

LOBES.

GARDNER's twelve
fish a pair of four inch
espect with the twelve
an entirely new and

Lightning Source UK Ltd.
Milton Keynes UK
UKHW011158051118
331792UK00006B/1067/P